Exploratory Data Analysis: An Introduction to Data Analysis Using SAS® Enterprise Guide

A Data-Based Approach

Exploratory Data Analysis: An Introduction to Data Analysis Using SAS® Enterprise Guide

Course Notes was developed by Data Services Online.

Copyright © 2007 Data Services Online. All rights reserved. Printed in the United States of America. No part of this publication may be reproduced, stored in a retrieval system, or transmitted, in any form or by any means, electronic, mechanical, photocopying, or otherwise, without the prior written permission of the publisher, Data Services Online.

ISBN 978-1-4357-0542-5

Table of Contents

Chapter 1. Introduction to Data Analysis 1

    1.1. Introduction 3

    1.2. Data Example Using Faculty Workload 4

    1.3. Data Example Using Student Survey 39

    1.4. MEPS Data 49

    1.5. Exercises 55

Chapter 2. Data Visualization with Kernel Density Estimation 56

    2.1. Introduction 58

    2.2. Kernel Density Estimation 60

    2.3. Faculty Workload Data 73

    2.4. Exercise 90

    2.5. Appendix 91

Chapter 3. Theory of Hypothesis Testing 93

    3.1. Introduction 95

        Example of Hypothesis Definition: 95

    3.2. Hypothesis Parameters 97

        Type I error 97

        Type II error 100

        Effect Size 102

        Sample Size 104

        The Problem of Generalizations 105

        Central Limit Theorem 106

3. Basics of Hypothesis Testing..107

3.4. Example of Faculty Workload...112

3.6. Sample Problem 125

3.7. Exercises...129

Chapter 4. Types of Statistical Analysis..130

4.1. Introduction 132

4.2. Data Mining...134

4.3. Cost-Effectiveness Analysis 135

4.4. Analysis of Variance 136

4.5. Processing Quality-of-Life Information 138

    Chi-Square Analysis 138

4.6. Examples 142

4.7. Exercises 157

Chapter 5. Basics of Inference for Categorical Data 158

5.1. Introduction 159

5.2. Data Summary..160

5.3. Discussion of the Problem...165

5.4. Chi Square Analysis 180

5.5. Logistic Regression 199

5.6. Exercises...212

Chapter 6. Linear Regression (Analysis of Variance)...213

6.1. Introduction 215

6.2. Linear Models in Enterprise Guide 216

6.3. Exercises..248

Chapter 7. Additional Examples from Student Papers..249

7.1. Example 1 of Logistic Regression..251

7.2. Example 2-Linear Models 259

Chapter 1. Introduction to Data Analysis

1.1. Introduction 3

1.2. Data Example Using Faculty Workload 4

1.3. Data Example Using Student Survey 39

1.4. Exercises 55

1.1. Introduction

The primary purpose of data analysis is to examine a set of data. The data can be relatively simple, or it can be very complex. There are two main approaches to data analysis. The first is *exploration*. We don't really know what information is in the data, and we use different statistical techniques to gain useful knowledge about the data. The second approach is *inference*. We have some idea of what is in the data, and we create a hypothesis in order to test our ideas.

We can think of the data as providing a narrative. They tell a story. We use our techniques of data analysis to extract the story from the data. Then, we write the narrative so that other people will be able to understand the story. Because the story is specific to the data set we are examining, it is difficult to simply develop techniques of data analysis. The techniques that are necessary for one dataset may not be important for another dataset. For this reason, it is important to learn a variety of different techniques. The most difficult decisions become those of choosing which techniques to use. The more techniques you know, the better you will become at data analysis.

Because you will need to write the narrative, it is also necessary to develop writing skills. There are certain forms that must be observed when writing a technical paper, and your papers must follow these forms. Different disciplines have different forms, so you need to learn a general method of writing.

Basic data exploration will be discussed in this chapter, with a specific type of data exploration given in Chapter 2. Chapter 3 will discuss the basics of statistical inference. Chapter 4 will examine elementary methods of statistical inference; chapter 5 gives a basic introduction to the exploration of categorical data; chapter 6 examines continuous data.

1.2. Data Example Using Faculty Workload

Because data must be examined in context, we will provide some information concerning the datasets we will be using here. The data set contains the workload assignments for university faculty in one academic department for a three-year period. There are a total of 14 variables and 71 observations. There are a number of reasons to examine the data: to determine if employee resources are used efficiently and to determine whether there are longitudinal shifts in workload assignments that might impact overall productivity.

One of the first things we want to do is to look at the variables, and the observations in the data. To do this, we use SAS Enterprise Guide. We start by defining a new project.

Display 1.1. Opening Screen for Enterprise Guide

Next, we open the datafile, faculty workload.

Display 1.2. Open Dataset in Enterprise Guide

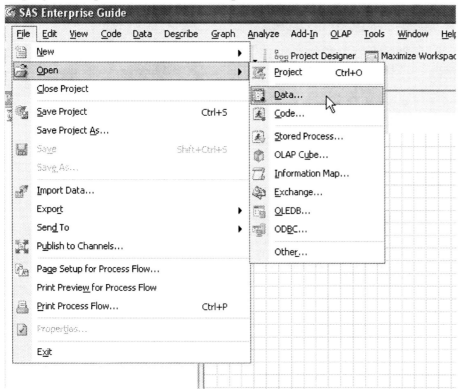

We want sheet one of the Excel spreadsheet and to use the option of conversion to a SAS dataset.

Display 1.3. Screen to Import Excel Spreadsheet

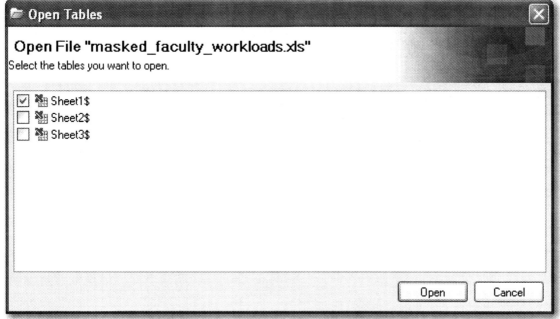

Display 1.4. Screen to Choose Type of Dataset

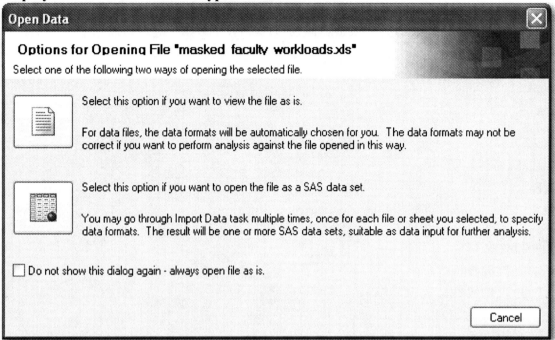

The first line of the spreadsheet contains the variable names, so we check the first box and then click run.

Display 1.5. Screen to Define Variable Headers

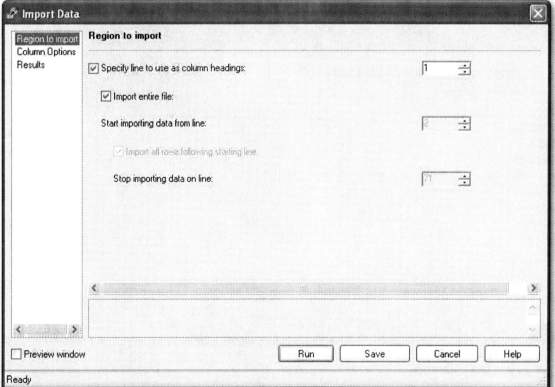

The dataset looks like the one in Display 1.6.

Display 1.6. Data Spreadsheet

	Faculty	Year	% Instruction	% Courses	Number of Courses	# Calculus Courses	Large
1	1	2003-2004	40	40	3	0	
2	1	2002-2003	50	50	4	0	
3	1	2001-2002	66	66	4	2	
4	2	2003-2004	50	50	4	2	
5	2	2002-2003	30	30	3	0	
6	2	2001-2002	38	30	3	0	
7	3	2003-2004	50	48	4	1	
8	3	2002-2003	53	50	4	0	
9	3	2001-2002	54	50	4	0	
10	4	2003-2004	58	50	4	0	
11	4	2002-2003	58	50	4	2	
12	4	2001-2002	60	50	4	2	
13	5	2003-2004	54	38	3	0	
14	5	2002-2003	58	48	4	0	
15	5	2001-2002	54	50	4	1	
16	6	2003-2004	0	0	0	0	
17	6	2002-2003	52	50	4	0	
18	6	2001-2002	52	50	4	0	
19	7	2003-2004	30	30	2	0	
20	7	2002-2003	10	10	1	0	
21	7	2001-2002	15	15	1	0	
22	8	2001-2002	56	56	4	2	
23	9	2001-2002	66	66	4	3	
24	10	2003-2004	25	25	2	1	
25	10	2002-2003	25	25	2	0	
26	10	2001-2002	25	25	2	1	

To examine the data, we want to get a list of the variables, and also some summary information.

There are three different kinds of data available: interval, ordinal, and nominal. Nominal often uses numeric codes to represent categories such as race and gender. Ordinal consist mostly of rankings. Interval values are numbers that can be used in numerical operations. We want to find the means and variances of interval variables; frequency counts of nominal and ordinal variables. In the dataset here, there are four nominal variables: faculty, year, rank, and sabbatical. Although the faculty variable appears to be numeric, these are codes substituting for names. Therefore, we first collect frequency counts of these nominal fields. We use the Describe menu.

Display 1.7. Menu for Procedures to Describe Data

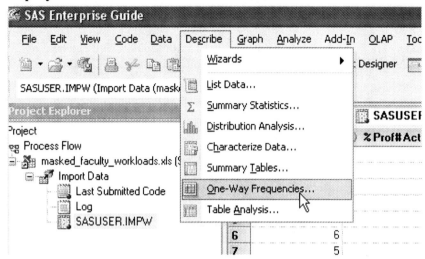

Display 1.8. Screen to Compute Variable Frequencies

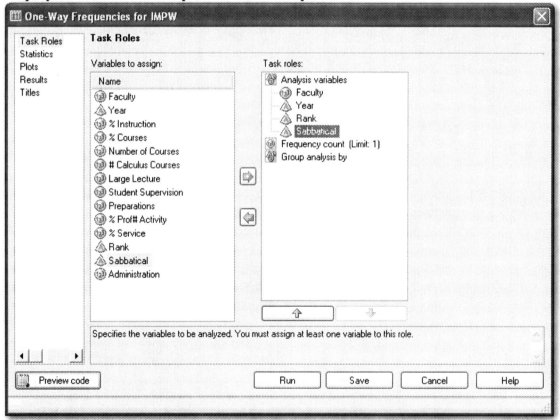

We get the following results shown in Tables 1.1-1.3.

Table 1.1. Number of Observations by Year

Year	Year			
Year	Frequency	Percent	Cumulative Frequency	Cumulative Percent
2001-2002	24	33.80	24	33.80
2002-2003	24	33.80	48	67.61
2003-2004	23	32.39	71	100.00

Table 1.2. Number of Observations by Faculty Rank

Rank	Rank			
Rank	Frequency	Percent	Cumulative Frequency	Cumulative Percent
Assistant	11	15.49	11	15.49
Associate	29	40.85	40	56.34
Professor	31	43.66	71	100.00

Table 1.3. Number of Sabbaticals

Sabbatical	Sabbatical			
Sabbatical	Frequency	Percent	Cumulative Frequency	Cumulative Percent
Fall	5	7.04	5	7.04
None	59	83.10	64	90.14
Spring	4	5.63	68	95.77
Year	3	4.23	71	100.00

Because the frequency count for each faculty member is 2 or 3, we don't reproduce it here. Otherwise, it indicates that there were 9 half-year sabbaticals and 3 full-year sabbaticals for the three years of data presented here. The numbers also indicate that there are more professors than associate and assistant professors.

For the interval variables, the mean is defined as the average. The standard deviation is defined as the amount of variability. Usually, any data falling outside of two standard deviations of the mean are considered to be outliers. Technically, the standard deviation is equal to the sum of the squared difference between each point in the data and the mean, divided by one less than the number of data points. For our data, we use the summary statistics (Display 1.9).

Display 1.9. Menu for Summary Statistics

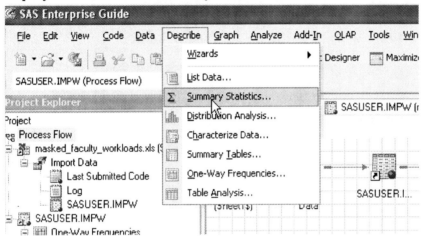

Display 1.10. Screen for Summary Statistics

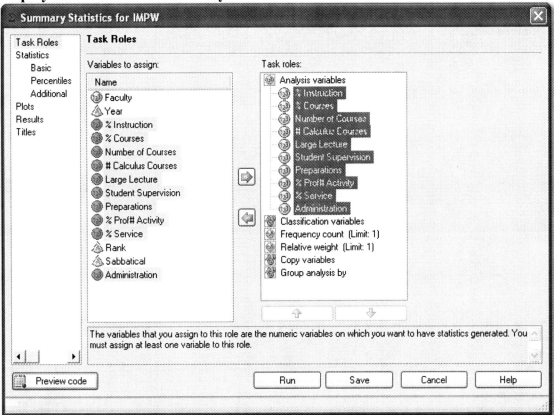

Then run the summary statistics (Table 1.4).

Table 1.4. Summary Statistics Results

Variable	Mean	Std Dev	Minimum	Maximum	N
% Instruction	41.6056338	16.9203840	0	66.0000000	71
% Courses	38.8732394	16.6286582	0	66.0000000	71
Number of Courses	3.0985915	1.2438986	0	5.0000000	71
# Calculus Courses	0.8732394	0.9400544	0	3.0000000	71
Large Lecture	0.5070423	0.8084065	0	2.0000000	71
Student Supervision	1.3098592	2.6327148	0	12.0000000	71
Preparations	0.7323944	2.0767952	0	10.0000000	71
% Prof# Activity	38.8591549	20.5581376	10.0000000	100.0000000	71
% Service	19.5352113	15.5488456	0	70.0000000	71
Administration	8.3802817	15.8964203	0	60.0000000	71

The results indicate that the average percent for courses is slightly less than the average percent for instruction, indicating that there are additional instructional responsibilities that are not part of teaching courses. The average time spent on professional activity is almost 40%; however, the standard deviation is considerable. There is at least one faculty member with a research workload of 100%. Domain knowledge tells us that this faculty member is on sabbatical.

The average number of students supervised is 1.3; the average number of calculus courses is almost one. Because these numbers can also be considered ordinal, we can compute frequency counts of them as well.

Display 1.11. Screen for Frequency Counts for Ordinal Variables

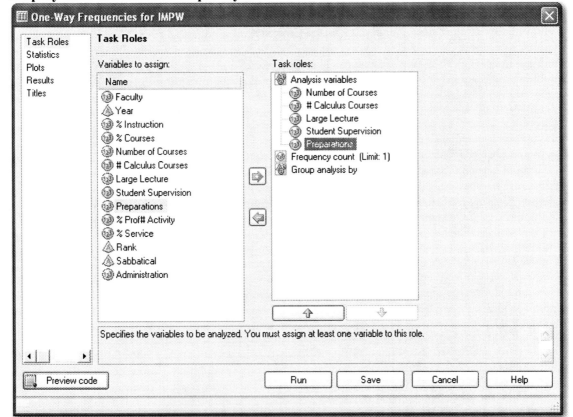

Table 1.5 shows that most of the faculty teach four courses; however, quite a few teach only 2 or 3.

Table 1.5. Frequency Counts for Number of Courses

Number of Courses				
Number of Courses	Frequency	Percent	Cumulative Frequency	Cumulative Percent
0	4	5.63	4	5.63
1	4	5.63	8	11.27
2	13	18.31	21	29.58
3	12	16.90	33	46.48
4	36	50.70	69	97.18
5	2	2.82	71	100.00

While the average number of calculus courses is almost one, it is clear that quite a few faculty do not teach calculus while others teach it frequently (Table 1.6).

Table 1.6. Frequency Counts for Number of Calculus Courses

# Calculus Courses				
# Calculus Courses	Frequency	Percent	Cumulative Frequency	Cumulative Percent
0	33	46.48	33	46.48
1	17	23.94	50	70.42
2	18	25.35	68	95.77
3	3	4.23	71	100.00

Similarly, some faculty are more likely to teach large lecture classes compared to others (Table 1.7).

Table 1.7. Frequency Counts for Number of Large Lecture Courses

Large Lecture				
Large Lecture	Frequency	Percent	Cumulative Frequency	Cumulative Percent
0	49	69.01	49	69.01
1	8	11.27	57	80.28
2	14	19.72	71	100.00

A few faculty members supervise many students; others supervise none.

Table 1.8. Frequency Counts for Number of Students Supervised

Student Supervision				
Student Supervision	Frequency	Percent	Cumulative Frequency	Cumulative Percent
0	49	69.01	49	69.01
2	11	15.49	60	84.51
3	2	2.82	62	87.32
4	1	1.41	63	88.73
5	4	5.63	67	94.37
8	1	1.41	68	95.77
10	1	1.41	69	97.18
11	1	1.41	70	98.59
12	1	1.41	71	100.00

Similarly, most faculty are not preparing new courses; others have considerable effort in new course preparations.

Table 1.9. Frequency Counts for Number of New Course Preparations

Preparations				
Preparations	Frequency	Percent	Cumulative Frequency	Cumulative Percent
0	61	85.92	61	85.92
2	1	1.41	62	87.32
3	2	2.82	64	90.14
4	2	2.82	66	92.96
5	2	2.82	68	95.77
6	1	1.41	69	97.18
10	2	2.82	71	100.00

Another way of looking at data is to break the observations down into various categories. For example, we can look at the variables with respect to year, and to faculty rank to see if there are differences. For the interval workload variables, we get the following by year, and then by rank.

Display 1.12. Screen for Summary Statistics of Continuous Variables by Year

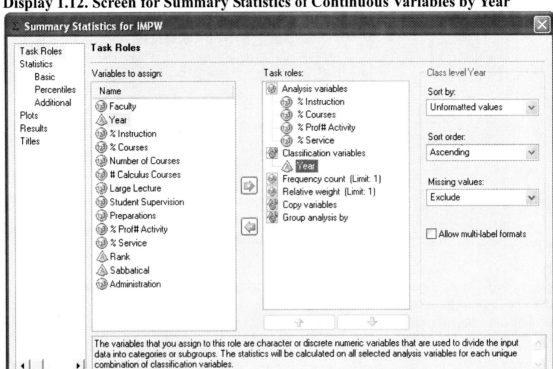

Table 1.10. Summary Statistics by Year

Year	N Obs	Variable	Label	Mean	Std Dev	Minimum	Maximum	N
2001-2002	24	% Instruction	% Instruction	48.4166667	16.0891988	10.0000000	66.0000000	24
		% Courses	% Courses	45.5833333	15.5309415	10.0000000	66.0000000	24
		% Prof# Activity	% Prof# Activity	32.5833333	13.5515655	15.0000000	67.0000000	24
		% Service	% Service	19.0000000	11.5645636	3.0000000	55.0000000	24
2002-2003	24	% Instruction	% Instruction	37.3333333	18.5815866	0	60.0000000	24
		% Courses	% Courses	35.1250000	18.2454580	0	55.0000000	24
		% Prof# Activity	% Prof# Activity	44.5833333	24.5868764	15.0000000	100.0000000	24
		% Service	% Service	18.0833333	16.9036057	0	70.0000000	24
2003-2004	23	% Instruction	% Instruction	38.9565217	14.1950000	0	58.0000000	23
		% Courses	% Courses	35.7826087	14.3240311	0	54.0000000	23
		% Prof# Activity	% Prof# Activity	39.4347826	21.0169328	10.0000000	80.0000000	23
		% Service	% Service	21.6086957	17.9892370	0	60.0000000	23

Notice that there is a considerable decrease in the percentage of instruction, and in the percentage of courses. The decrease is almost 10% between 2001-2002 and 2002-2003. There is a corresponding increase in the percentage of professional activity (research). One of the questions that needs to be answered using domain knowledge, is what happened to cause this reduction?

Table 1.11. Summary Statistics by Rank

Rank	N Obs	Variable	Label	Mean	Std Dev	Minimum	Maximum	N
Assistant	11	% Instruction	% Instruction	46.3636364	10.7449777	30.0000000	66.0000000	11
		% Courses	% Courses	44.5454545	10.8845178	30.0000000	64.0000000	11
		% Prof# Activity	% Prof# Activity	42.0909091	13.8307957	22.0000000	60.0000000	11
		% Service	% Service	11.5454545	5.3172105	6.0000000	23.0000000	11
Associate	29	% Instruction	% Instruction	49.4827586	15.3985637	0	66.0000000	29
		% Courses	% Courses	46.8620690	14.7955692	0	66.0000000	29
		% Prof# Activity	% Prof# Activity	36.0000000	17.2461176	20.0000000	100.0000000	29
		% Service	% Service	14.5172414	6.8275738	0	31.0000000	29
Professor	31	% Instruction	% Instruction	32.5483871	15.9746439	0	61.0000000	31
		% Courses	% Courses	29.3870968	15.3268336	0	60.0000000	31
		% Prof# Activity	% Prof# Activity	40.3870968	25.1126494	10.0000000	100.0000000	31
		% Service	% Service	27.0645161	20.1211920	0	70.0000000	31

Full professors have a much lower teaching load compared to assistant and association professors. Their research percentage is about the same, but there is also a considerable difference in the service workload.

We can use a table analysis to compare year and rank to the nominal and ordinal variables. We show some of the tables here.

Display 1.13. Menu for Table Analysis

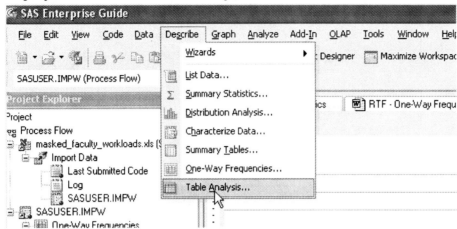

Display 1.14. Screen for Table Analysis

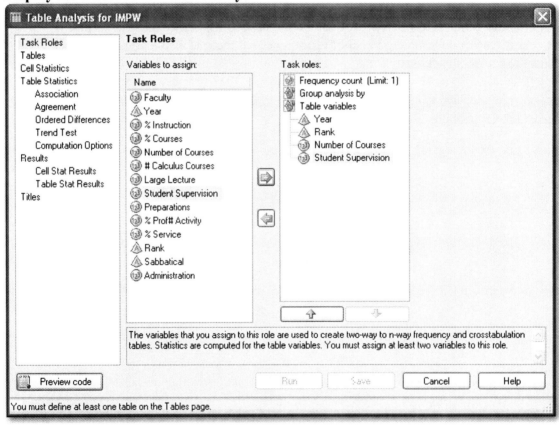

We define the tables by dragging the variable to indicate the row and column of the table. We can create different combinations of tables using the same screen. The default is to show only column percentages; we check the row percentages to compute those as well.

Display 1.15. Screen to Create Tables

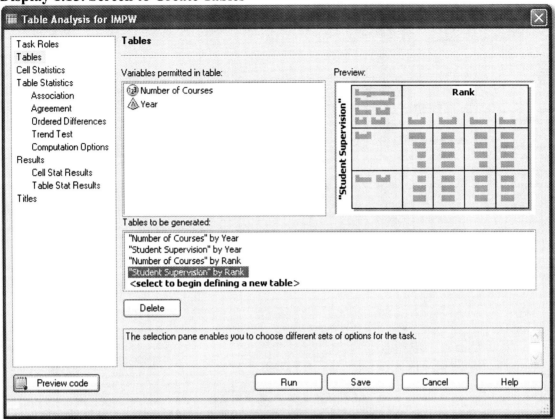

Display 1.16. Screen to Add Row Percentages to Table

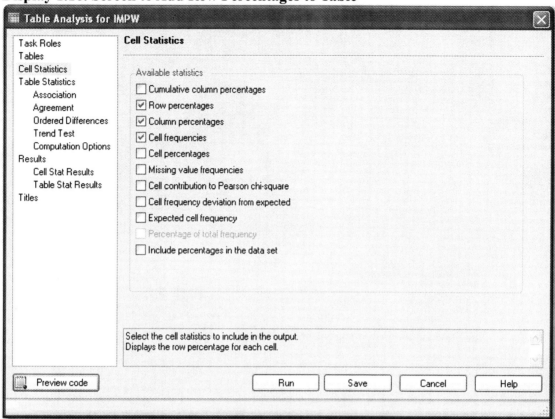

Table 1.12. Table of Number of Courses by Year

Number of Courses Frequency Row Pct Col Pct	Year(Year) 2001-2002	2002-2003	2003-2004	Total
0	0 0.00 0.00	3 75.00 12.50	1 25.00 4.35	4
1	2 50.00 8.33	1 25.00 4.17	1 25.00 4.35	4
2	2 15.38 8.33	4 30.77 16.67	7 53.85 30.43	13
3	3 25.00 12.50	5 41.67 20.83	4 33.33 17.39	12
4	16 44.44 66.67	10 27.78 41.67	10 27.78 43.48	36
5	1 50.00 4.17	1 50.00 4.17	0 0.00 0.00	2
Total	24	24	23	71

This frequency table indicates that there was a shift downward from sixteen faculty teaching four courses in 2001-2002, to only ten teaching four courses in the other two years.

Table 1.13. Table Analysis of Student Supervision by Year

Student Supervision Frequency Row Pct Col Pct	Year(Year) 2001-2002	Year(Year) 2002-2003	Year(Year) 2003-2004	Total
Table of Student Supervision by Year				
0	16 32.65 66.67	19 38.78 79.17	14 28.57 60.87	49
2	5 45.45 20.83	1 9.09 4.17	5 45.45 21.74	11
3	0 0.00 0.00	1 50.00 4.17	1 50.00 4.35	2
4	1 100.00 4.17	0 0.00 0.00	0 0.00 0.00	1
5	1 25.00 4.17	1 25.00 4.17	2 50.00 8.70	4
8	0 0.00 0.00	1 100.00 4.17	0 0.00 0.00	1
10	0 0.00 0.00	1 100.00 4.17	0 0.00 0.00	1
11	1 100.00 4.17	0 0.00 0.00	0 0.00 0.00	1
12	0 0.00 0.00	0 0.00 0.00	1 100.00 4.35	1
Total	24	24	23	71

There does not appear to be much of a shift in the number of students supervised from one year to another.

Table 1.14. Table Analysis of Number of Courses by Faculty Rank

Table of Number of Courses by Rank				
Number of Courses	**Rank(Rank)**			**Total**
Frequency Row Pct Col Pct	Assistant	Associate	Professor	
0	0 0.00 0.00	2 50.00 6.90	2 50.00 6.45	4
1	0 0.00 0.00	0 0.00 0.00	4 100.00 12.90	4
2	0 0.00 0.00	1 7.69 3.45	12 92.31 38.71	13
3	4 33.33 36.36	1 8.33 3.45	7 58.33 22.58	12
4	7 19.44 63.64	25 69.44 86.21	4 11.11 12.90	36
5	0 0.00 0.00	0 0.00 0.00	2 100.00 6.45	2
Total	11	29	31	71

By rank, only 13% of full professors teach 4 courses (and another 6.5% teach 5) compared to 86% of associate professors and 64% of assistants. Fully 39% of full professors teach 2 courses.

Table 1.15. Table Analysis of Student Supervision by Faculty Rank

Table of Student Supervision by Rank				
Student Supervision	Rank(Rank)			
Frequency Row Pct Col Pct	Assistant	Associate	Professor	Total
0	7 14.29 63.64	18 36.73 62.07	24 48.98 77.42	49
2	4 36.36 36.36	4 36.36 13.79	3 27.27 9.68	11
3	0 0.00 0.00	2 100.00 6.90	0 0.00 0.00	2
4	0 0.00 0.00	1 100.00 3.45	0 0.00 0.00	1
5	0 0.00 0.00	3 75.00 10.34	1 25.00 3.23	4
8	0 0.00 0.00	1 100.00 3.45	0 0.00 0.00	1
10	0 0.00 0.00	0 0.00 0.00	1 100.00 3.23	1
11	0 0.00 0.00	0 0.00 0.00	1 100.00 3.23	1
12	0 0.00 0.00	0 0.00 0.00	1 100.00 3.23	1
Total	11	29	31	71

More assistant and associate professors supervise students compared to full professors. The diagram in Enterprise Guide that we have used to perform the above steps is shown in Display 1.17. By saving the diagram, we save all of the analysis steps, and we can refer back to them as needed.

Display 1.17. Project Diagram for Data Summaries

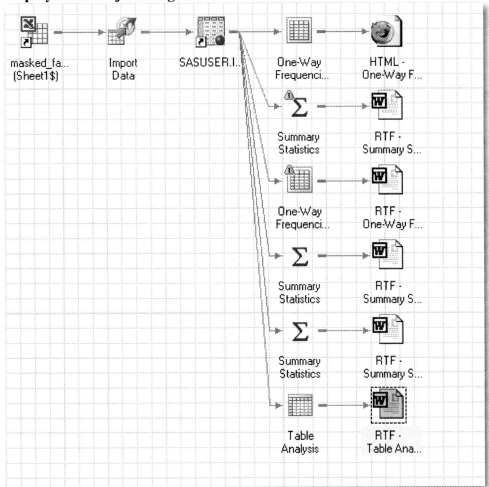

Thus far, we have used only summary statistics and frequency counts. Sometimes, a graph can give a better representation of the data. We begin with some basic bar charts. Notice that there are many options available.

Display 1.18. Graphics Menu

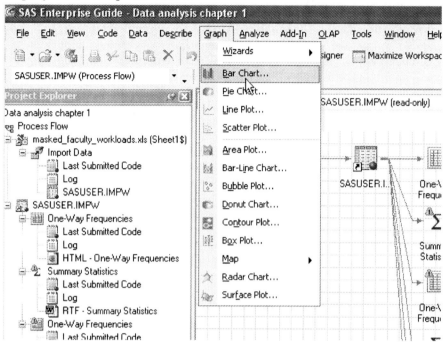

Display 1.19. Screen for Bar Graphs

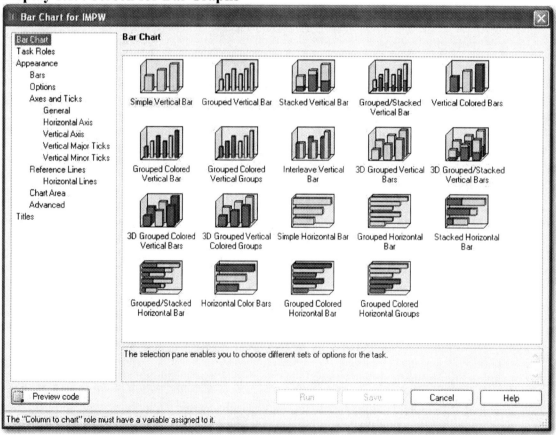

In this example, we will choose the grouped colored vertical bar so that we can compare by year and by rank.

Display 1.20. Screen for Grouped, Colored Vertical Bar

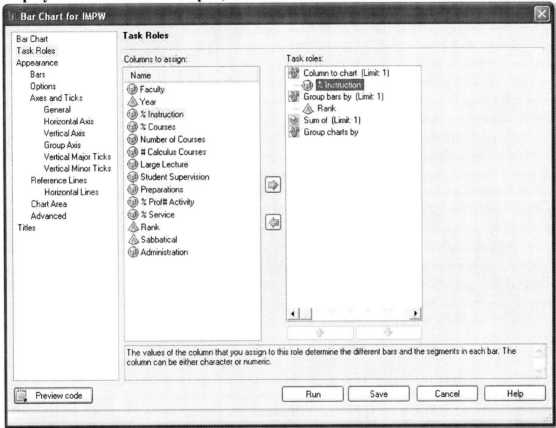

Figure 1.1. Bar Graph of Instruction by Rank

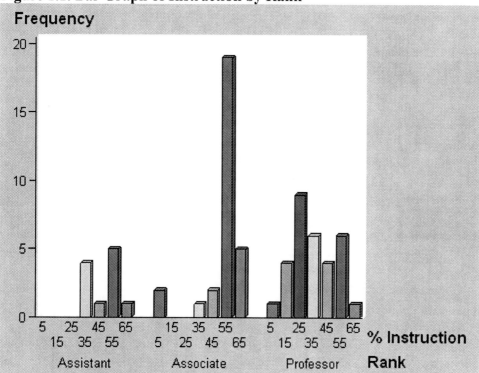

The bars are clearly spread out more for the full professors; almost all of the associate professors have the same percentage of instruction. The bar graph can very clearly show differences in variability between groups.

Figure 1.2. Bar Graph of Instruction by Year

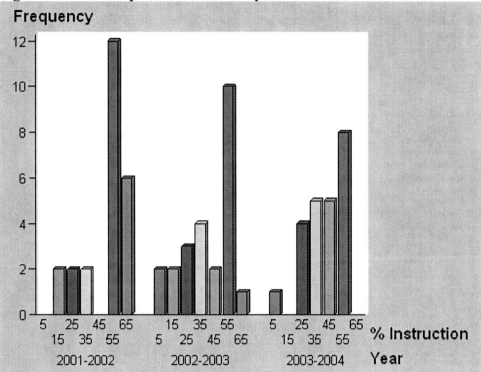

This bar graph shows the shift downward. The last bar for 65% is much higher for the year 2001-2002 compared to the year 2003-2004.

We can also use a grouped and stacked bar to compare the percentage of instruction by both year and rank.

Display 1.21. Screen for Grouped, Stacked Bar

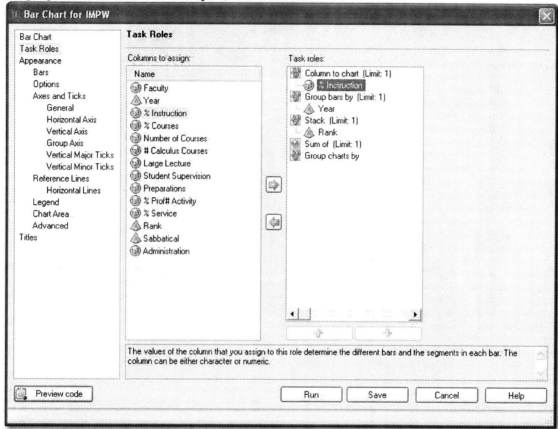

Figure 1.3. Grouped, Stacked Bar Graph of Instruction by Rank

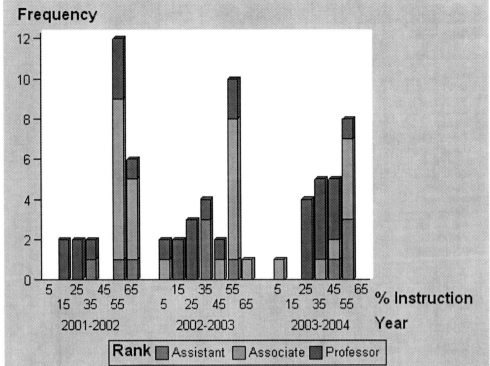

This graph shows clearly that full professors dominate the lower percentages for each year. We can reverse the position of rank and year. This graph shows more of the variability, but less of the contrast by rank. It is not as useful.

Display 1.22. Display for Grouped, Stacked Bar Graph by Year

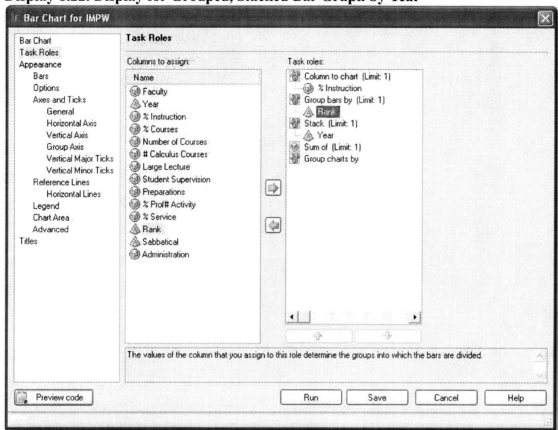

Figure 1.4. Comparison of Instruction, Rank, and Year

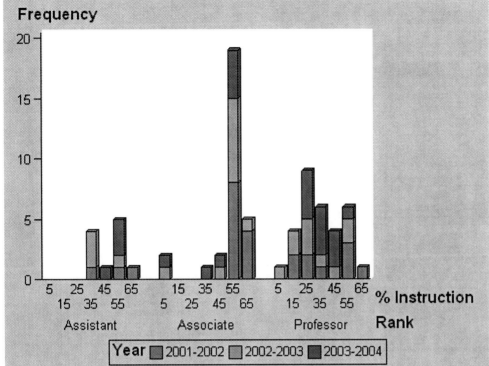

Another graph to consider, especially for nominal and ordinal data is a pie chart.

Display 1.23. Screen for Pie Chart

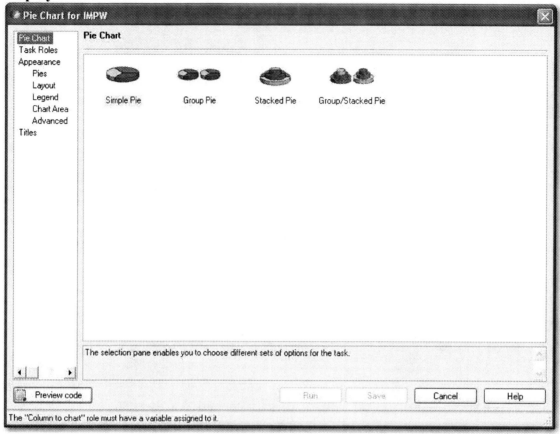

We will first consider a grouped pie chart.

Display 1.24. Screen for Pie Chart Details

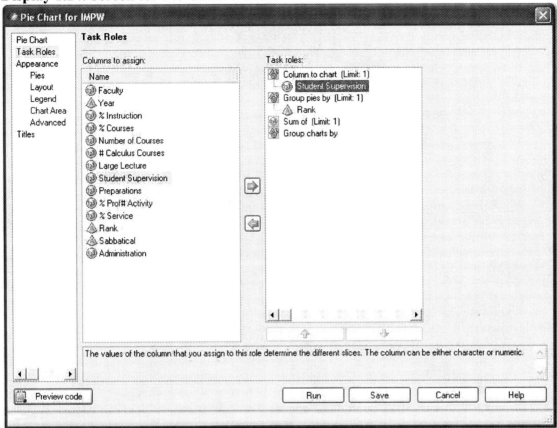

Figure 1.5. Pie Charts by Rank

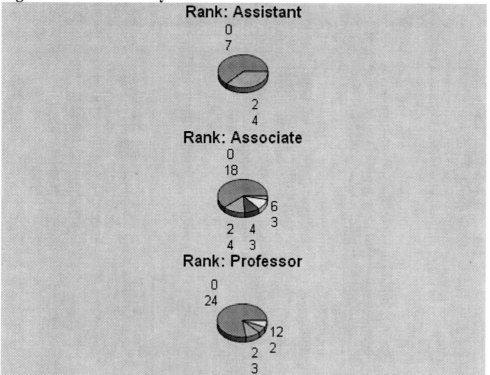

It shows that the majority of faculty are supervising no students. It we switch to a stacked graph rather than a grouped graph, we get the following (Figure 1.6).

Figure 1.6. Stacked Pie Chart

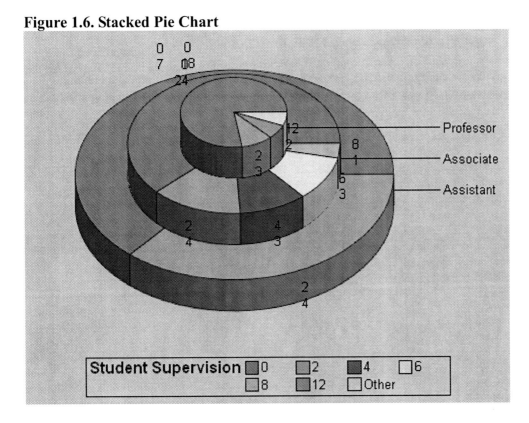

It depicts the same information, but in a slightly different way. Again, using a grouped/stacked pie chart, we can consider both rank and year.

Figure 1.7. Stacked, Grouped Pie Chart

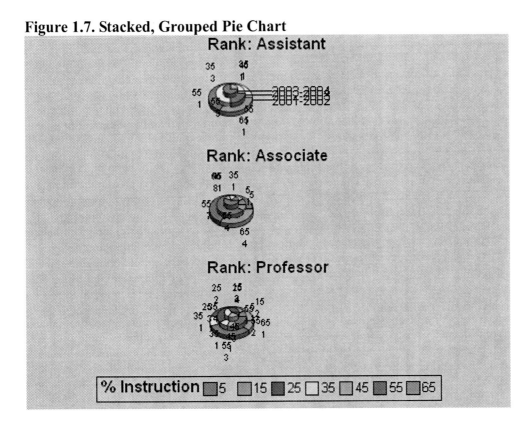

The graph is starting to get a little crowded and is probably not the best to use.

Given the information here, you can probably write a narrative of the data. Consider what additional information you might want to extract using summary statistics, and basic graphs.

1.3. Data Example Using Student Survey

During the 2001-2002 academic year, we surveyed all students in mathematics courses numbered 200 or above. The questions are given below:

      Course_____   Section_____
1. The areas of mathematics which interest me are (Check all that apply):
   _____Pure mathematics
   _____Applied mathematics
   _____Abstract algebra
   _____Number theory
   _____Topology
   _____Real analysis
   _____Discrete mathematics
   _____Differential equations
   _____Actuarial science
   _____Probability
   _____Statistics
2. I am currently enrolled in the following mathematics classes:
   _____.
3. The mathematics courses I would like to take in the spring and fall semesters are:
   _____.
4. Maple was used in the following classes:
   _____.
5. Other computer software or calculators were used in the following classes:
   _____.
6. I want [more, less (circle one)] use of the computer software in class.
7. The computer lab taught in calculus was [very useful, somewhat useful, of little use, totally useless, I am just starting calculus].

The computer lab in calculus was [very organized and well coordinated with class material, somewhat organized and partially coordinated with class material].

I would prefer that the applied mathematics courses focus on [concepts, computation]. Currently, they focus on [concepts, computation].

(agree, disagree) I would like to get a BS degree in mathematics.

(agree, disagree) I could not get the classes I needed to fulfill an applications area for the BS degree.

(agree, disagree) The courses I couldn't get for my applications area were in the mathematics department.

(agree, disagree) The courses I couldn't get for my applications area were in other departments.

I am working toward a BA degree and the reason is_____.

I am no longer working toward a mathematics major because _____.

I do not want to major in mathematics because _____.

(agree, disagree) A mathematics major is prepared only to teach after graduation.

(agree, disagree) I would like to do a co-op in an area business.

In comparison with other students in mathematics, I think my skills and knowledge are [very weak, weak, average, strong, very strong].

I usually spend _____ hours per week studying mathematics.

Course instructors [rarely, occasionally, frequently, usually, always] expect me to know things that were never previously covered in class.

Course instructors [rarely, occasionally, frequently, usually, always] expect me to know things that I remember seeing in a previous class but have forgotten.

There is [way too much, too much, about right, too little, way too little] material taught in mathematics courses.

The workload required is [way too much, too much, about right, too little, way too little].

(agree, disagree) I understand the links between the different mathematics courses in the program.

(agree, disagree) I can see the links between course content and current research in mathematics.

A course designed to teach how to do mathematical proofs is [of great value, of some value, of little value, of no value] to me.

My work habits and study methods [are inadequate, are adequate, are more than adequate] to really succeed in studying mathematics.

In the dataset, the variables are defined by column:

Variable Name	Corresponding Question in Survey
Student_Type	No question
CourseLevel	Condensed from initial question on course
Course	Initial question on course
Section	Initial question on section
Pure	Question 1
Applied	Question 1
AbstractAlgebra	Question 1
NumberTheory	Question 1
Topology	Question 1
RealAnalysis	Question 1
DiscreteMathematics	Question 1
DifferentialEquations	Question 1
ActuarialScience	Question 1
Probability	Question 1
Statistics	Question 1
_enrolled	Question 2
_wanttoenroll	Count of courses in Question 3
_maple	Question 4
_othersoftware	Question 5
Usecomputer	Question 6
Calculusmaple	Question 7
Qualitylab	Question 8
Appliedmathdoes	Question 9, first response
Appliedmathshould	Question 9, second response
BS	Question 10
Notgetclasses	Question 11
Notgetmath	Question 12

Variable Name	Corresponding Question in Survey
BA	Question 14
Quitmajor	Question 15
Notmajor	Question 16
Teachmajor	Question 17
Intern	Question 18
Comparestudents	Question 19
Hours	Question 20
Expectknownever	Question 21
Expectknowprevious	Question 22
Toomuch	Question 23
Links	Question 25
Linksresearch	Question 26
Proofs	Question 27
Workhabits	Question 28
Coursestaken	No question
Coursesnotscheduled	No question
FormNumber	No question-identifier

The rest of this section gives a narrative of the data.

A total of 190 student survey forms were collected: 130 in courses numbered 200-300 and 60 in courses numbered 400 and above. The distribution by course level is below:

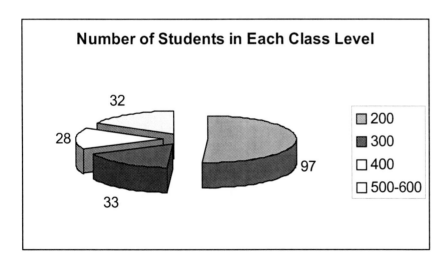

Of that number, the following percentages of students expressed interest in pure and applied mathematics (students could check both categories if they desired):

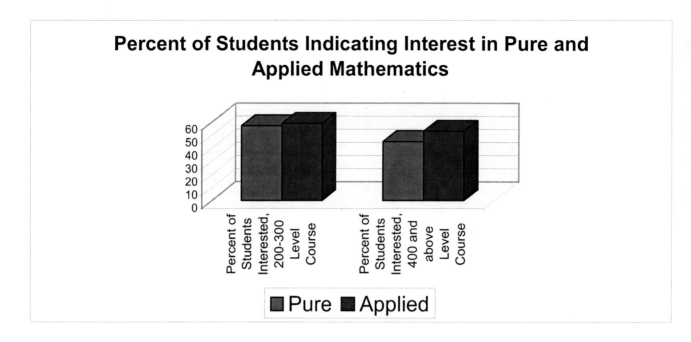

Note that in general categories, the students are relatively split between the two categories. However, when they are broken down by individual mathematical discipline, there are differences:

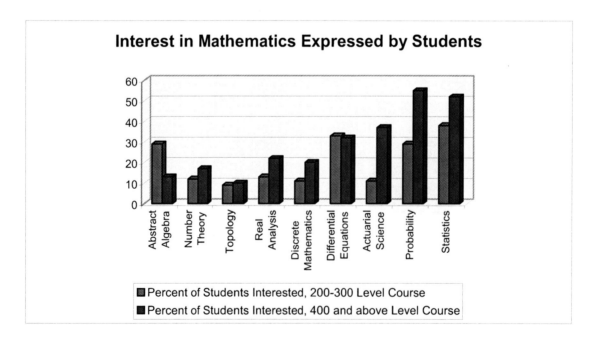

Students express greater interest in the applied mathematics sequences, particularly probability, statistics, and actuarial science. This interest increases for students in higher level mathematics courses.

On the average, students in courses 200-300 had taken one mathematics course previously to the one enrolled and expressed an interest in one and a half additional courses (on average). For students in upper level courses, this average increases to 2. The overall distribution is shown for students in lower level courses and in upper level courses.

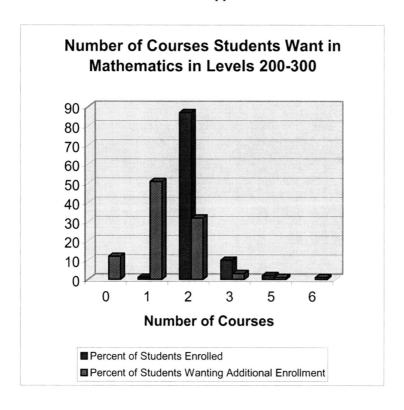

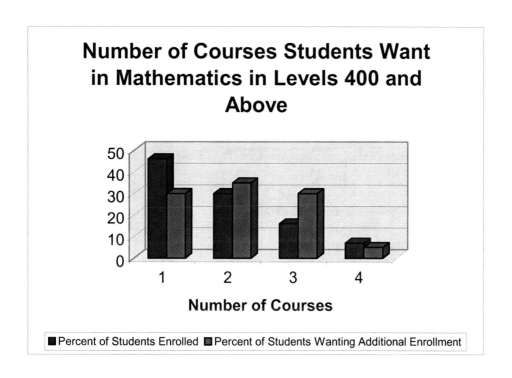

In order to succeed in mathematics courses, students need to invest in study time. Therefore, students were asked to estimate the amount of time they studied. Overall, the distribution is as follows:

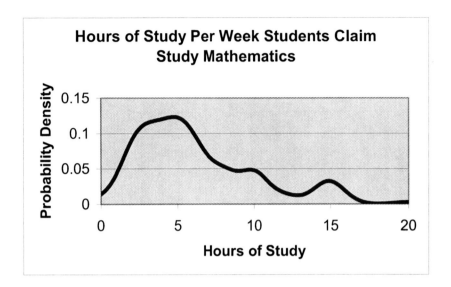

For a 3 hour course, students should study 9-10 hours outside of class and for a 5 hour course, they should study approximately 15 hours per week. Students on the average are investing half that amount of time. If the study times are compared by course level:

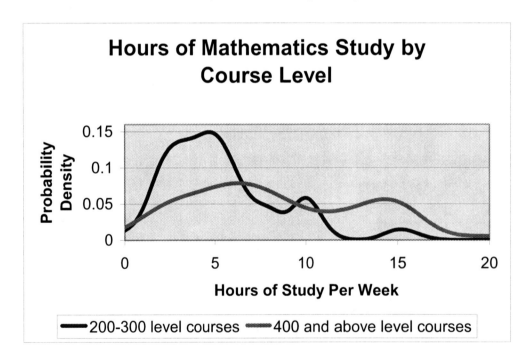

It is noticeable that students in 200-300 level courses are investing less than 5 hours per week. As these are largely 5 hour courses, they are studying approximately 1/3 of the time needed. Students in upper level courses have increased the number of hours but these are spread across more than one mathematics course.

The Department of Mathematics has recently added a Maple lab to all calculus courses. Statistical software has been required in all statistics courses for quite some time. It is an essential part of the modeling sequence. Student opinions concerning the use of computers were also solicited:

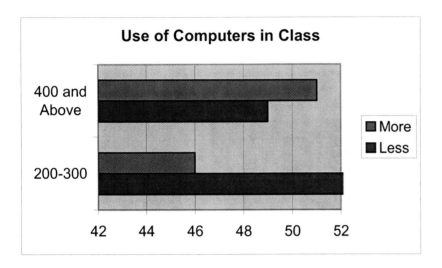

Note that the majority of students in 200-300 level courses (with the required Maple lab) want less use of the computer; this trend reverses in upper level courses. Students were also asked specifically about the use of Maple in calculus:

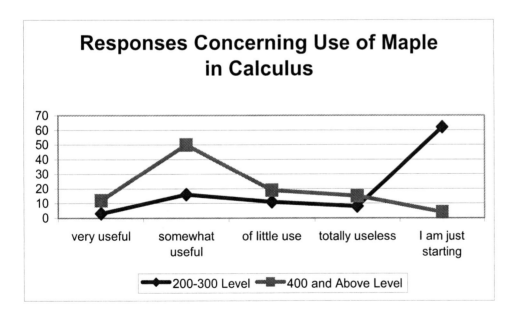

Again, note that students in upper level courses found the Maple to be useful with a substantial minority (25%) believing that the calculus lab had little use. Students in lower level courses were evenly divided in their estimation of the use of the lab.

Restricting attention to those students with experience in Maple indicates that a sizeable proportion (50%) find the Maple lab of little use:

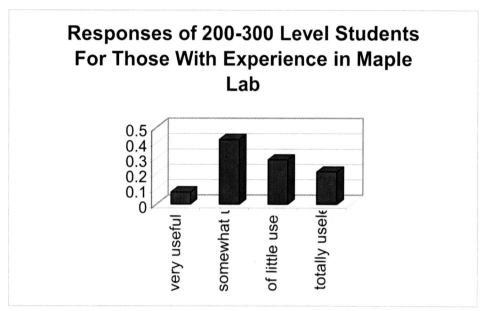

Since the Department of Mathematics is contemplating the development of a Master's and PhD program in industrial and applied mathematics, students were asked about their experiences in applied courses as well we their interests. Applied courses can focus on computation, manipulating numbers and variables or they can focus on concepts using computational or statistical software to perform the computations. Students preferred that these courses focus on concepts but found that they primarily focused on computations:

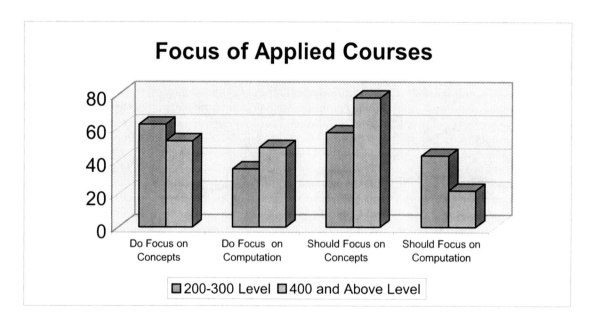

Students in the upper level courses have a strong opinion that the courses focus on computations but should focus on concept. With the use of Maple and other computer software, computations can be performed with the aid of the computer so that it is possible to focus on concepts.

Students also have an interest in internships ranging from 52% in beginning level courses, increasing to 64% in upper level courses:

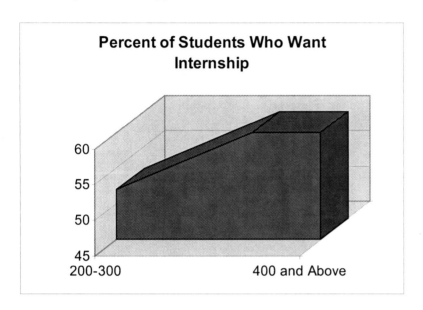

Students were asked if they had difficulties enrolling in mathematics courses. In the 200-300 level courses, a small percentage (13%) did have difficulty:

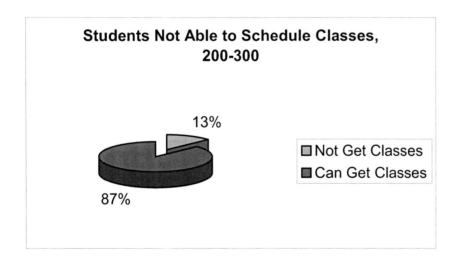

However, this percentage climbed to 34% for students in upper level classes:

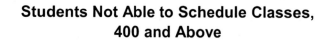

1.4. MEPS Data

Many of the datasets used in this text are from the Medical Expenditure Panel Survey (http://www.meps.ahrq.gov/mepsweb/). Data were collected (up through 2004 as of December, 2006) concerning a cohort of individuals and households. The datasets available for download include those given in Figure 1.8.

Figure 1.8. Data Files Available Through MEPS

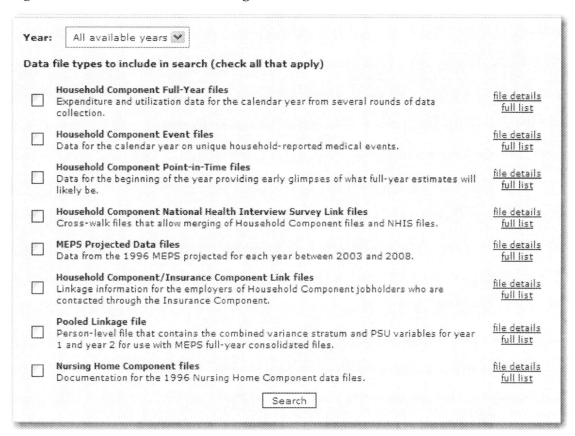

By specifying a year, we can find all datasets available for that year. For the year 2003, the files are given in Figure 1.9.

Figure 1.9. Data Files for the Year 2003

LINK_2003HC/NHIS	2003 MEPS/2002 & 2001 NHIS Link File		2003	NHIS Link File
HC-086	MEPS Panel 8 Longitudinal Weight File		2003-2004	Household Full Year File
HC-080	MEPS Panel 7 Longitudinal Weight File	X	2002-2003	Household Full Year File
HC-079	2003 Full Year Consolidated Data File		2003	Household Full Year File
HC-078	2003 Medical Conditions File		2003	Household Full Year File
HC-077I	Appendix to MEPS 2003 Event Files		2003	Household Event File
HC-077H	2003 Home Health File		2003	Household Event File
HC-077G	2003 Office-Based Medical Provider Visits File		2003	Household Event File
HC-077F	2003 Outpatient Department Visits		2003	Household Event File
HC-077E	2003 Emergency Room Visits File		2003	Household Event File
HC-077D	2003 Hospital Inpatient Stays File		2003	Household Event File
HC-077C	2003 Other Medical Expenses		2003	Household Event File
HC-077B	2003 Dental Visits		2003	Household Event File
HC-077A	2003 Prescribed Medicines File		2003	Household Event File
HC-076	2003 Person Round Plan Public Use File	X	2003	Household Full Year File
HC-074	2003 Jobs File		2003	Household Full Year File
HC-073	2003 Full Year Population Characteristics (HC-073 replaced by HC-079)		2003	Household Full Year File
HC-064	2003 P7R3/P8R1 Population Characteristics		2003	Household Point-in-Time File
HC-036BRR	MEPS 1996-2004 Replicates for Calculating Variances File		1996-2004	Pooled Linkage File
HC-036	MEPS 1996-2004 Pooled Estimation Linkage File		1996-2004	Pooled Linkage File

The list of files contains virtually every encounter with healthcare providers. All of the data files are keyed to individual and to household identifiers. Once the user identifies a specific data file for download, instructions to do so are given (Figure 1.10).

Figure 1.10. Download Data Files

Documentation	File type
Documentation	PDF (349 KB) / HTML
Codebook	PDF (185 KB) / HTML
SAS Programming Statements	ASCII format (51 KB)
SPSS Programming Statements	ASCII format (20 KB)

Data	File type*
Data File, ASCII format	ZIP (3.7 MB) / EXE
Data File, SAS transport format	ZIP (5.3 MB) / EXE

*Right click on the data file link, then select Save Target As... or Save Link As... to download the file.

The files exist either in standard tab-delimited text format (ASCII) or in SAS transport files. We recommend that the SAS transport files are used. While the documentation is extensive, the SAS code to import these files is fairly straightforward:

```
LIBNAME PUFLIB 'C:\MEPS';
FILENAME IN1 'C:\MEPS\H77G.SSP';

PROC XCOPY IN=IN1 OUT=PUFLIB IMPORT;
RUN;
```

The code assumes that the SAS transport file is copied to the C:\MEPS directory on the local computer's hard drive. For now, the datasets are located in the c:\project_directory. In SAS format.

We will focus on the 2004 Prescription Medication Datafile, HC-085A. The variables in the dataset are listed in Figure 1.11.

Figure 1.11. Variable Definitions

Name	Start	End	Description
CLMOMFLG	368	368	CHGE/PYMNT, Rx CLAIM FILING, OMTYPE STAT
DUID	1	5	DWELLING UNIT ID
DUPERSID	9	16	PERSON ID (DUID + PID)
INPCFLG	369	369	PID HAS AT LEAST 1 RECORD IN PC
LINKIDX	32	43	ID FOR LINKAGE TO COND/OTH EVENT FILES
PCIMPFLG	367	367	TYPE OF HC TO PC PRESCRIPTION MATCH
PERWT04F	519	530	FINAL PERSON LEVEL WEIGHT, 2004
PHARTP1	352	353	TYPE OF PHARMACY PROV - 1ST
PHARTP2	354	355	TYPE OF PHARMACY PROV - 2ND
PHARTP3	356	357	TYPE OF PHARMACY PROV - 3RD
PHARTP4	358	359	TYPE OF PHARMACY PROV - 4TH
PHARTP5	360	361	TYPE OF PHARMACY PROV - 5TH
PHARTP6	362	363	TYPE OF PHARMACY PROV - 6TH
PHARTP7	364	365	TYPE OF PHARMACY PROV - 7TH
PID	6	8	PERSON NUMBER
PREGCAT	389	390	MULTUM PREGNANCY CATEGORY
PURCHRD	44	44	ROUND Rx/PRESCR MED OBTAINED/PURCHASED
RXBEGDD	45	46	DAY PERSON STARTED TAKING MEDICINE
RXBEGMM	47	48	MONTH PERSON STARTED TAKING MEDICINE
RXBEGYRX	49	52	YEAR PERSON STARTED TAKING MEDICINE
RXCCC1X	380	382	MODIFIED CLINICAL CLASS CODE
RXCCC2X	383	385	MODIFIED CLINICAL CLASS CODE
RXCCC3X	386	388	MODIFIED CLINICAL CLASS CODE
RXFLG	366	366	NDC IMPUTATION SOURCE ON PC DONOR REC
RXFORM	152	201	FORM OF Rx/PRESCRIBED MEDICINE (IMPUTED)
RXFRMUNT	202	251	UNIT OF MEAS FORM Rx/PRESC MED (IMPUTED)
RXHHNAME	103	132	HC REPORTED MEDICATION NAME
RXICD1X	371	373	3 DIGIT ICD-9 CONDITION CODE
RXICD2X	374	376	3 DIGIT ICD-9 CONDITION CODE
RXICD3X	377	379	3 DIGIT ICD-9 CONDITION CODE
RXMD04X	447	453	AMOUNT PAID, MEDICAID (IMPUTED)
RXMR04X	440	446	AMOUNT PAID, MEDICARE (IMPUTED)

RXNAME	53	102	MEDICATION NAME (IMPUTED)
RXNDC	133	143	NATIONAL DRUG CODE (IMPUTED)
RXOF04X	475	480	AMOUNT PAID, OTHER FEDERAL (IMPUTED)
RXOR04X	499	505	AMOUNT PAID, OTHER PRIVATE (IMPUTED)
RXOT04X	493	498	AMOUNT PAID, OTHER INSURANCE (IMPUTED)
RXOU04X	506	511	AMOUNT PAID, OTHER PUBLIC (IMPUTED)
RXPV04X	454	460	AMOUNT PAID, PRIVATE INSURANCE (IMPUTED)
RXQUANTY	144	151	QUANTITY OF Rx/PRESCR MED (IMPUTED)
RXRECIDX	17	31	UNIQUE Rx/PRESCRIBED MEDICINE IDENTIFIER
RXSF04X	433	439	AMOUNT PAID, SELF OR FAMILY (IMPUTED)
RXSL04X	481	486	AMOUNT PAID, STATE & LOCAL GOV (IMPUTED)
RXSTRENG	252	301	STRENGTH OF Rx/PRESCR MED DOSE (IMPUTED)
RXSTRUNT	302	351	UNIT OF MEAS STRENGTH OF Rx (IMPUTED)
RXTR04X	468	474	AMOUNT PAID, TRICARE (IMPUTED)
RXVA04X	461	467	AMOUNT PAID, VETERANS (IMPUTED)
RXWC04X	487	492	AMOUNT PAID, WORKERS COMP (IMPUTED)
RXXP04X	512	518	SUM OF PAYMENTS RXSF04X-RXOU04X(IMPUTED)
SAMPLE	370	370	HOUSEHLD RCVD FREE SAMPLE OF Rx IN ROUND
TC1	391	393	MULTUM THERAPEUTIC CLASS #1
TC1S1	394	396	MULTUM THERAPEUTIC SUB-CLASS #1 FOR TC1
TC1S1_1	397	399	MULTUM THERAPEUT SUB-SUB-CLASS FOR TC1S1
TC1S1_2	400	402	MULTUM THERAPEUT SUB-SUB-CLASS FOR TC1S1
TC1S2	403	405	MULTUM THERAPEUTIC SUB-CLASS #2 FOR TC1
TC1S2_1	406	408	MULTUM THERAPEUT SUB-SUB-CLASS FOR TC1S2
TC2	409	411	MULTUM THERAPEUTIC CLASS #2
TC2S1	412	414	MULTUM THERAPEUTIC SUB-CLASS #1 FOR TC2
TC2S1_1	415	417	MULTUM THERAPEUT SUB-SUB-CLASS FOR TC2S1
TC2S1_2	418	420	MULTUM THERAPEUT SUB-SUB-CLASS FOR TC2S1
TC2S2	421	423	MULTUM THERAPEUTIC SUB-CLASS #2 FOR TC2
TC3	424	426	MULTUM THERAPEUTIC CLASS #3
TC3S1	427	429	MULTUM THERAPEUTIC SUB-CLASS #1 FOR TC3
TC3S1_1	430	432	MULTUM THERAPEUT SUB-SUB-CLASS FOR TC3S1
VARPSU	534	534	VARIANCE ESTIMATION PSU, 2004
VARSTR	531	533	VARIANCE ESTIMATION STRATUM, 2004

With values generally represented as in Figure 1.12.

Figure 1.12. MEPS Codes

VALUE		DEFINITION
-1	INAPPLICABLE	Question was not asked due to skip pattern
-7	REFUSED	Question was asked and respondent refused to answer question
-8	DK	Question was asked and respondent did not know answer
-9	NOT ASCERTAINED	Interviewer did not record the data
-13	VALUE SUPPRESSED	Data suppressed
-14	NOT YET TAKEN/USED	Respondent answered that the medicine has not yet been used

1.5. Exercises

1. Given the analysis in section 2 concerning the faculty workload, write a narrative that explains the data. Duplicate this analysis using the Wordloads dataset.
2. Given the narrative in section 3 concerning the student survey, analyze the data using summary statistics and graphics to examine the data for a similar narrative.
3. Copy information on MEPS dataset. Pick one of the datasets and examine it using summary statistics and graphs.

Chapter 2. Data Visualization with Kernel Density Estimation

2.1. Introduction 58

2.2. Kernel Density Estimation 60

2.3. Faculty Workload Data 73

2.4. Exercise 90

2.5. Appendix 91

2.1. Introduction

In many analyses, it is common to assume that the data follow a bell-shaped curve. Most of you have seen this graphic (Figure 2.1).

Figure 2.1. The Bell-Shaped Curve

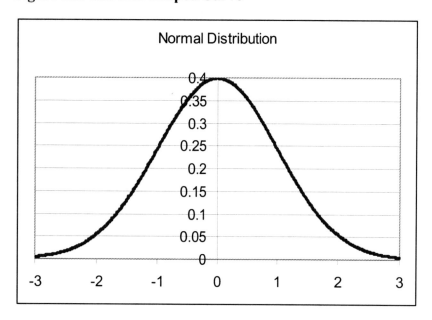

This assumption is not always valid. For example, consider the value of height. Since men and women have different average values, the distribution of the heights (assuming that men and women occur in equal numbers) is more likely to be bimodal (Figure 2.2).

Figure 2.2. Bimodal Curve

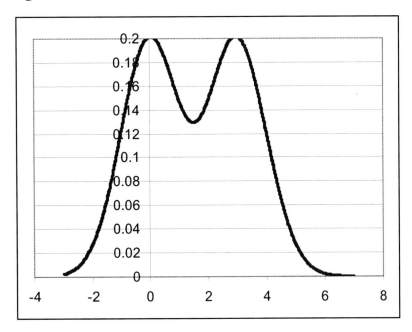

If we add children in to the distribution, the number of "bumps" will increase. The key word is "homogeneous". If the population is homogeneous, we can assume that it generally follows a normal distribution. If the population is heterogeneous, we must assume that it is multimodal in some way. Therefore, we need a way to estimate the distribution.

2.2. Kernel Density Estimation

While a bar chart can show the overall distribution of a population (Figure 2.3, Number of packyears for individuals entering a clinical trial), a kernel density estimation can smooth out the histogram (Figure 2.4). Although not specifically available in Enterprise Guide, it is in SAS/Stat and can be used with a SAS Code Node.

Figure 2.3. Histogram of Packyears

Figure 2.4. Kernel Density of Packyears Per Patient

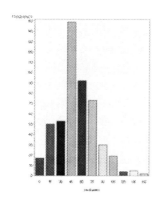
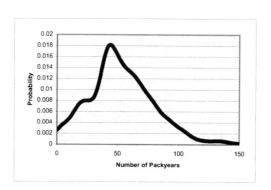

One of the advantages of using kernel density estimation is that different populations can be compared on the same graph; histogram comparisons must be made side by side (Figure 2.5).

Figure 2.5. Comparison of Packyears by Race

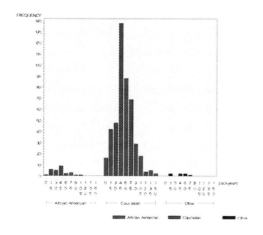
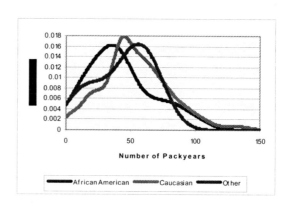

Because Caucasians represent most of the population, side-by-side comparisons just reflect the difference in population sizes rather than actual differences in proportions. Since kernel density estimators can be compared on the same graph, it can easily be seen that the kernel density for African Americans is to the left of that for Caucasians. It can then be concluded that African

Americans in the study population generally report that they smoke fewer cigarettes over a lifetime. Kernel density estimators also use all of the data instead of blocking it first as in a traditional bar graph. Therefore, kernel density estimators can provide a better fit of the data.

Unfortunately, kernel density is not an available option in Enterprise Guide and must be written in SAS code. The basic code is

```
Proc kde data=work.packyearsdataset;
Univar packyears / out=outkde;
Run;
```

In order to use kernel density estimation with Enterprise Guide, we will need to insert code into EG. This is done by using the SAS Code node. We access the code node through the menu (Display 2.1).

Display 2.1. Menu to Access SAS Code Node

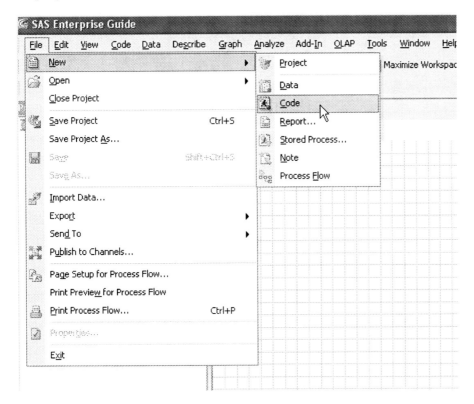

Display 2.2. Window to Enter SAS Code

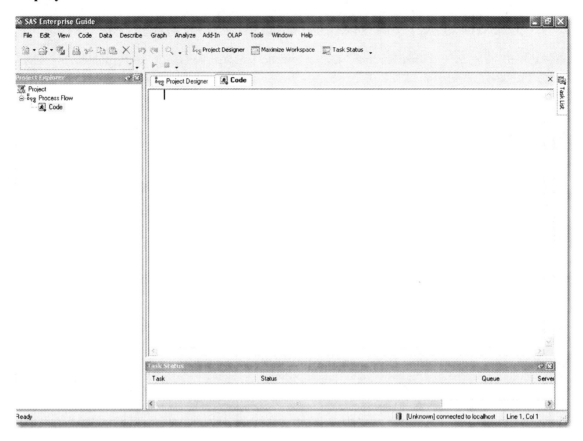

We just type in the code in the code node. When finished, just right click in the white space and click on Run on Local (Display 2.3).

Display 2.3. Run SAS Code

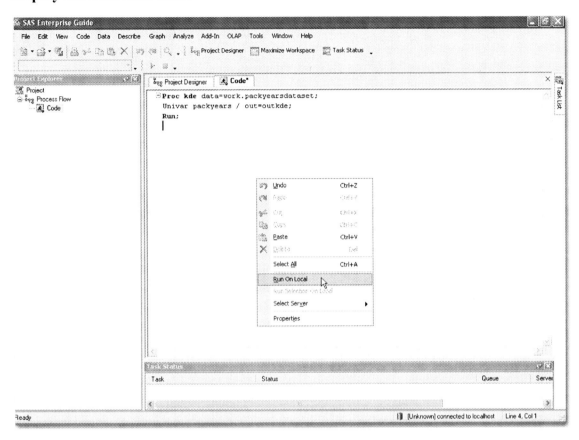

PROC KDE computes a data function containing a default 401 datapoints (Table 2.1). It must be stored as a SAS dataset (outkde).

Table 2. 1. First 10 (out of 401) rows of outkde

packyears	density	count
0	0.002551	7
0.375	0.002654	0
0.75	0.002755	0
1.125	0.002852	2
1.5	0.002946	0
1.875	0.003036	1
2.25	0.003123	0
2.625	0.003207	0
3	0.003288	0

The graph is computed by using a scatterplot with packyears as the x-variable and density as the y-variable. The count gives the number of actual values in the dataset that are equal to the corresponding x-value. A higher count total will compute a higher density value.

It is possible to fix the beginning and ending point of the procedure through the addition of lower and upper grid limits:

```
Proc kde data=work.packyearsdataset;
Univar packyears / gridl=0 gridu=150 out=outkde;
Run;
```

New to version 9, more than one Univar statement can be contained within Proc KDE. The above program will compute the density for packyears from the point 0 to the point of 150 packyears. To compare multiple populations, a by statement must be added:

```
Proc kde data=work.packyearsdataset;
Univar packyears / out=outkde;
By Race;
Run;
```

However, the by statement will only work if the data are first sorted by the same variable:

```
Proc sort data=work.packyearsdataset;
By Race;
Proc kde data=work.packyearsdataset;
Univar packyears out=outkde;
Run;
```

There is one more primary parameter that needs to be considered, the bandwidth. It controls the level of smoothing in the kernel density estimator. The smaller the bandwidth value, the more jagged the estimator (Figure 2.6).

Figure 2.6. Kernel density estimator at different bandwidth levels

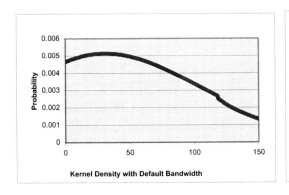
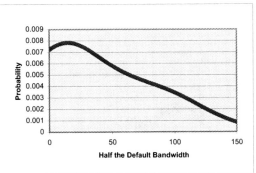
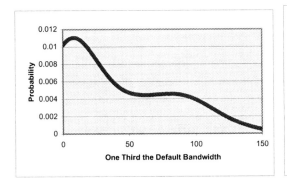
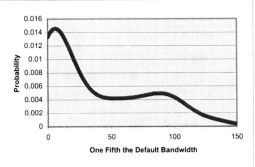

The default bandwidth gives an estimator that is over-smoothed. Using one third or one fifth of the default bandwidth gives a better representation of the data. There are four methods used by SAS to define a bandwidth (OS, SROT, SNR, SJPI). SJPI gives the default. To specify a bandwidth method, one statement is added to the SAS program:

```
Proc kde data=work.packyearsdataset;
Univar packyears / gridl=0 gridu=150 method=SROT out=outkde;
Run;
```

To use a multiple of the default bandwidth, you can specify:

```
Proc kde data=work.packyearsdataset;
Univar packyears / gridl=0 gridu=150 method=SROT bwm=.20 out=outkde;
Run;
```

Figure 2.7. gives a comparison of the bandwidth methods.

Figure 2.7. Comparison of Methods of Bandwidth Estimation

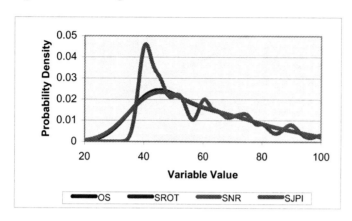

Note that method SJPI tends to give a kernel density estimator that is more jagged than the other three, which give estimators that are almost identical but perhaps too smooth.

Figure 2.8 uses the same data but a multiple of the bandwidth. For SJPI, the bandwidth multiplier is larger than 1, and gives a result smoother than the default. For the remaining three, the multiple is less than one, to give a result not quite as smooth.

Figure 2.8. Comparison of Methods of Bandwidth Estimation Using a Multiplier

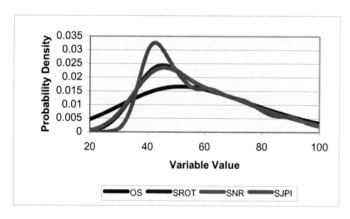

By using multipliers, the different methods used to compute the bandwidth start to converge in value.

Lung capacity is defined by a measure called the FEV/FVC ratio. A ratio of 100% is considered normal; a threshold of 70% indicates the presence of a lung disease. Consider the FEV/FVC ratio values collected along with packyears (Figure 2.9).

Figure 2.9. Comparison of two populations with same density but shift in values

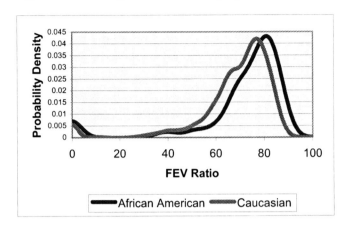

In this example, the two graphs are almost identical but one is shifted exactly to the right of the other with a difference of about 5%. Such a shift is nearly impossible to visualize using a histogram (Figure 2.10.)

Figure 2.10 Histogram comparing the 2 populations for FEV/FVC ratios

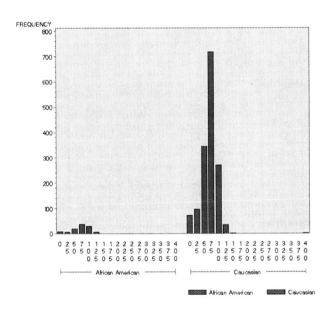

Several examples of kernel density estimators are given below, along with an explanation of the meaning of the curves.

Figure 2.11. Comparison of 6 different populations

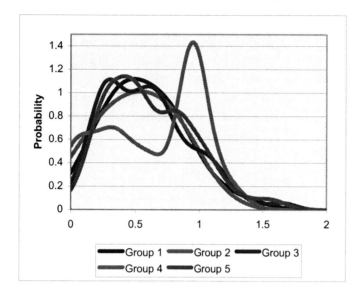

Groups 1-3 and 5 have relatively similar kernel density estimators, and there is little difference between the populations. Group 4, however, is considerably different from the other groups. Group 4 has a much lower probability of having a value less than 0.8, and a much greater probability of having a value that is greater than 1.

Figure 2.12. Comparison of Six Populations

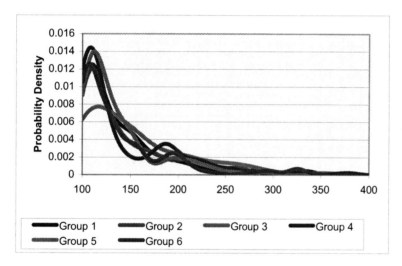

In this set of 6 populations, there is a clear ordering of the curves. For this particular value, it is desirable for the population value to stay less than 150, and a definition of disease if it is at 180 or greater. In this case, groups 4 and 5 are the least likely to have disease problems; group 3 has the greatest likelihood of disease.

Figure 2.13. Comparison of length of hospital stay by treatment drug

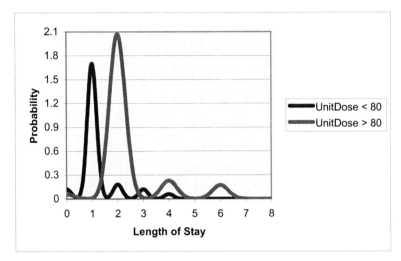

In this example, there is a clear separation between the two population groups. It is clear that patients receiving a higher dose of medication are generally likely to stay 2 days in the hospital compared to 1 day for those patients receiving a lower dose.

Although in previous versions of SAS, PROC KDE did not directly graph the density estimators, returning only a sufficient number of points to allow you to graph the estimators using SAS/Graph or exporting to a spreadsheet program to graph, the latest version has incorporated the capability. Version 9 will also allow you to graph multiple variables using the same procedure statement. However, although version 9 PROC KDE will provide a graph automatically, it will not work with the BY statement to put multiple populations on the same graph. Also, the ODS option must be on in order for the graphics to work. The following code will provide different univariate graphics (Figure 2.14-16).

```
ods html;
    ods graphics on;
proc kde data=sasuser.cabg;
univar pack_years / plots=density histogram histdensity;
run;
```

Figure 2.14. gives the histogram only.

Figure 2.14. ODE Output of PROC KDE for Plots=Histogram

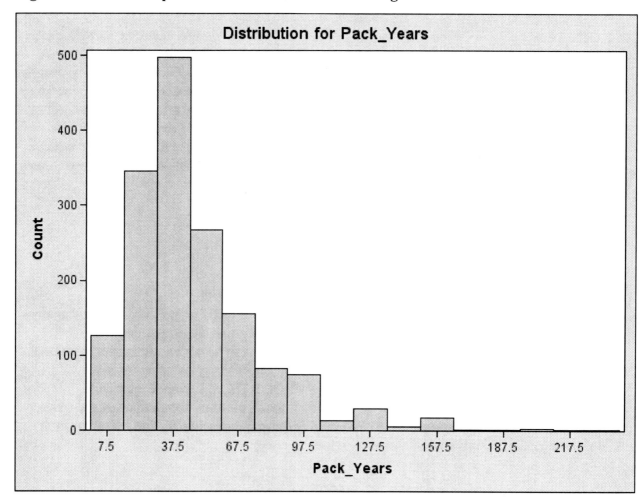

Figure 2.15 gives the Kernel Density Function.

Figure 2.15. ODE Output of PROC KDE for Plots=Density

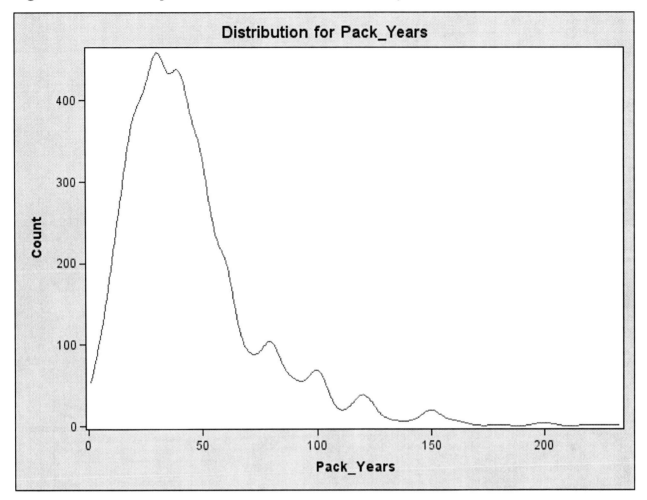

Figure 2.16 gives both the histogram and the density function.

Figure 2.16. ODE Output of PROC KDE for Plots=Histdensity

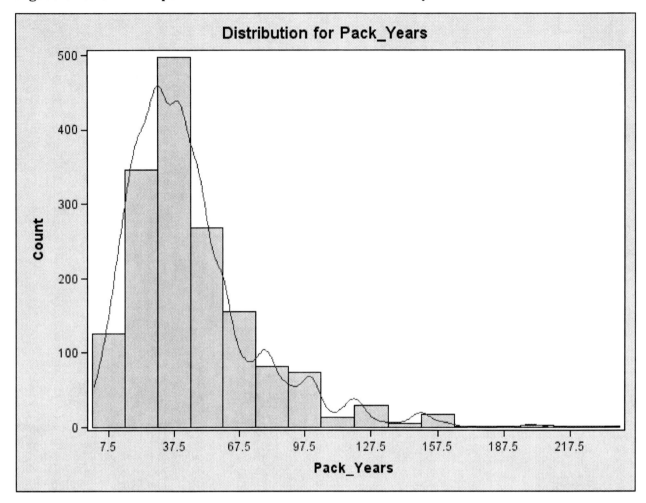

2.3. Faculty Workload Data

We will continue our study of the faculty workload data. In particular, we want to compare values by rank, and by year. In order to do this, we use the following code (Display 2.4).

First, we use the Sorting capability in Enterprise Guide to sort the data by year.

Display 2.4. Menu for Sorting Datasets

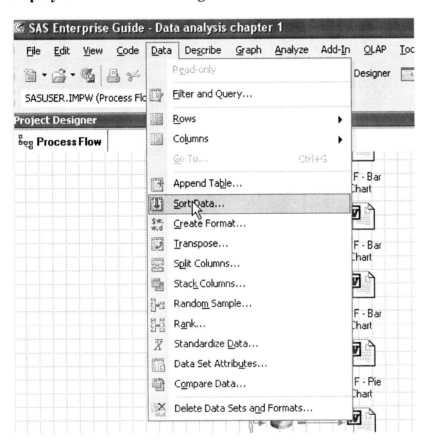

Display 2.5. Screen to Set Up Data Sort

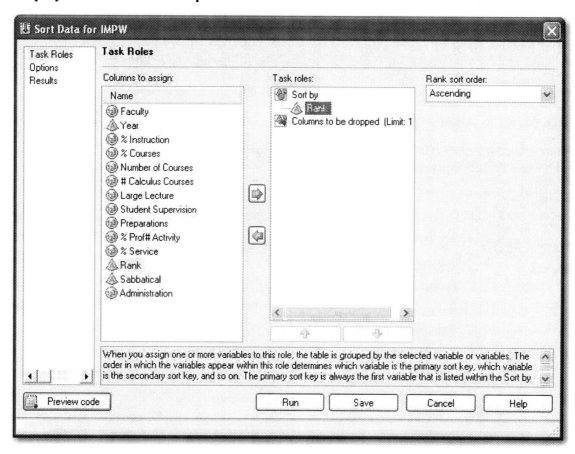

SAS assigns a default name to the sorted file.

Display 2.6. Dataset to Store Sorted Data

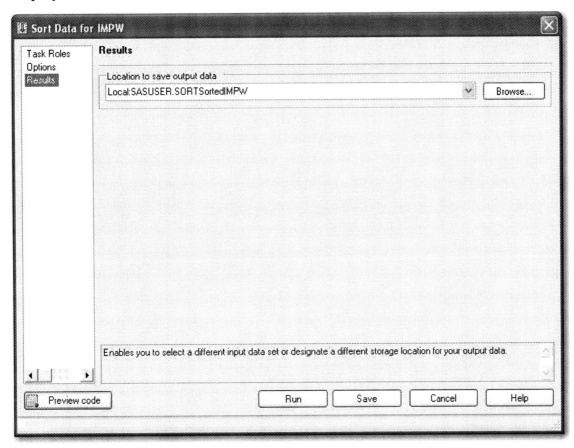

We want to change the name to something more meaningful, so we change it to the following (Display 2.7).

Display 2.7. Define Dataset Name

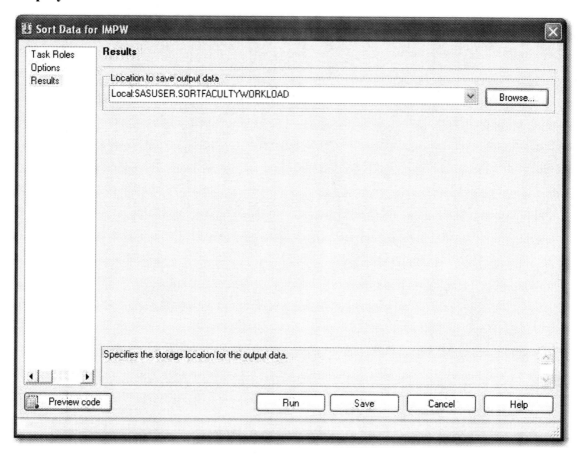

The next step is to "fix" the variable name so that the code node can read it. We go to the Data menu and uncheck the "read only" option. We click on the sorted datafile. Then we right-click on the variable name to go to the properties option.

Display 2.8. Menu to Rename Variables

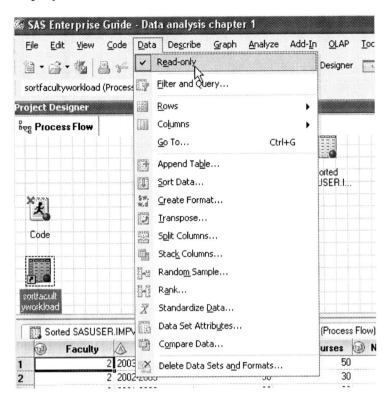

Display 2.9. Menu to Access Variable Properties

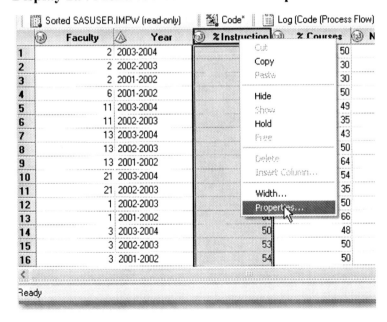

A text box comes up (Display 2.10).

Display 2.10. Screen to Change Variable Name

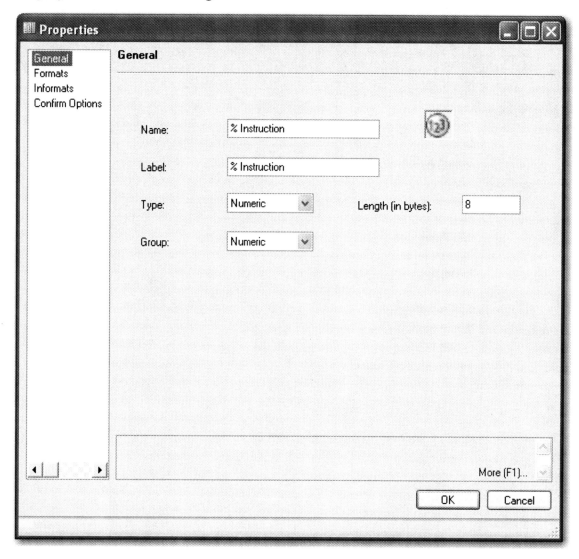

We change the name to 'Instruction'(Display 2.11).

Display 2.11. Change Name

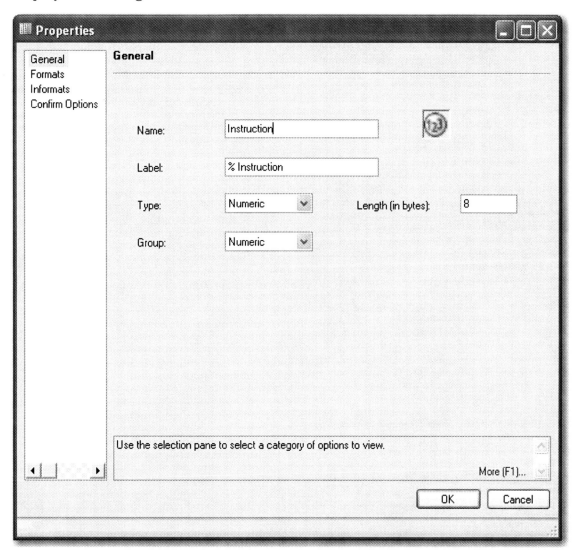

Then we re-check the 'read only' option. The reason for this change is that the code node will not recognize special characters in the variable names. This tends to cause confusion.

Display 2.12. Change to Read Only

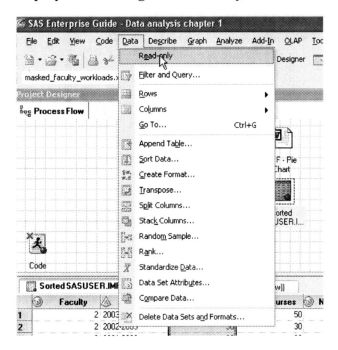

We then write the following code in a code node:

```
Proc kde data= sasuser.sortfacultyinstruction;
Univar Instruction/gridl=0 gridu=100
out=sasuser.kdeinstructionbyrank;
By rank;
Run;
```

The code node generates some output, and also a dataset (Display 2.13).

Display 2.13. Diagram for Code Node

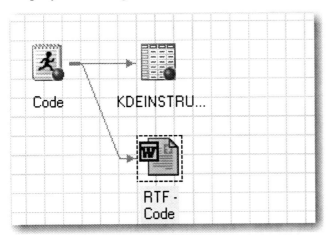

We click on the dataset and go to the line plot graph. The new dataset has the following variables (Display 2.14).

Display 2.14. KDE Output

	Rank	var	value	density	count
1	Assistant	Instruction	0	0	0
2	Assistant	Instruction	0.25	1.897354E-19	0
3	Assistant	Instruction	0.5	5.963112E-19	0
4	Assistant	Instruction	0.75	2.710505E-19	0
5	Assistant	Instruction	1	2.710505E-19	0
6	Assistant	Instruction	1.25	2.46656E-18	0
7	Assistant	Instruction	1.5	4.119968E-18	0
8	Assistant	Instruction	1.75	7.264155E-18	0
9	Assistant	Instruction	2	1.477225E-17	0
10	Assistant	Instruction	2.25	2.485533E-17	0
11	Assistant	Instruction	2.5	4.464202E-17	0
12	Assistant	Instruction	2.75	7.876729E-17	0
13	Assistant	Instruction	3	1.463673E-16	0
14	Assistant	Instruction	3.25	2.619432E-16	0
15	Assistant	Instruction	3.5	4.697035E-16	0
16	Assistant	Instruction	3.75	8.34402E-16	0
17	Assistant	Instruction	4	1.479936E-15	0

The 'value' variable gives the x-axis ranging from the lower grid value to the upper grid value. The 'density' variable gives the y-axis.

Display 2.15. Menu for Line Plot

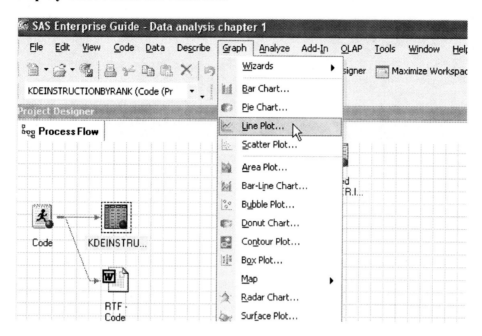

Because we are putting multiple groups on the same graph, we want to use the 'Multiple line plots' option. Otherwise, we would use the 'Line plot' option or the 'spline plot' option.

Display 2.16. Multiplot Selection

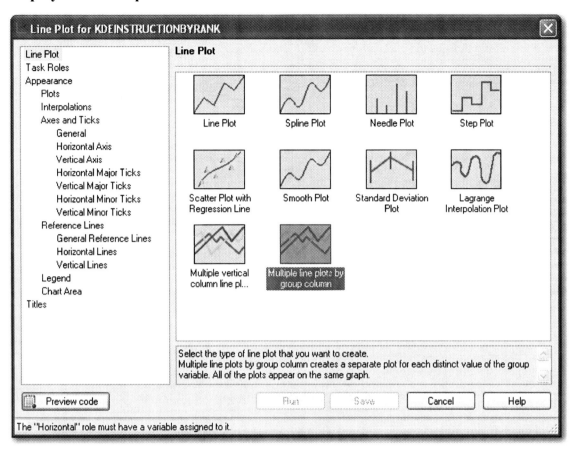

We assign the variables and then hit 'run'.

Display 2.17. Screen for Multiplot

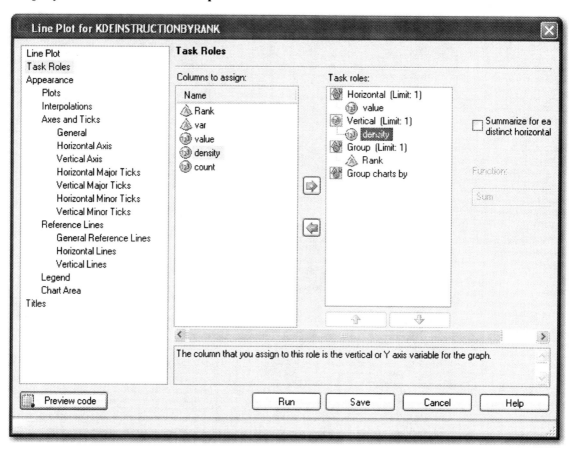

Figure 2.17. Kernel Density of Instruction by Rank

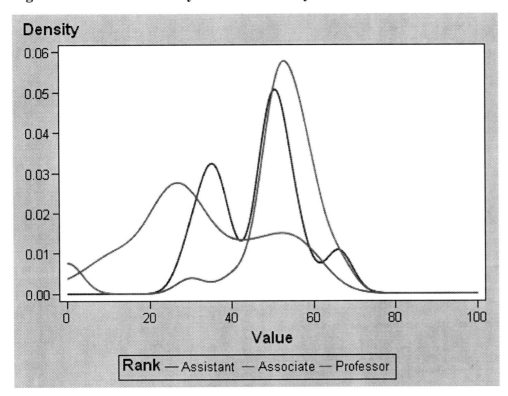

This graph allows us to compare the ranks more directly rather than to use the bar graph. It shows very clearly that the associate rank has the highest teaching load (it has the most rightward peak). The full professors have the least teaching load, with two peaks at 20% and 45%.

Figure 2.18. Bar Graph of Instruction by Rank

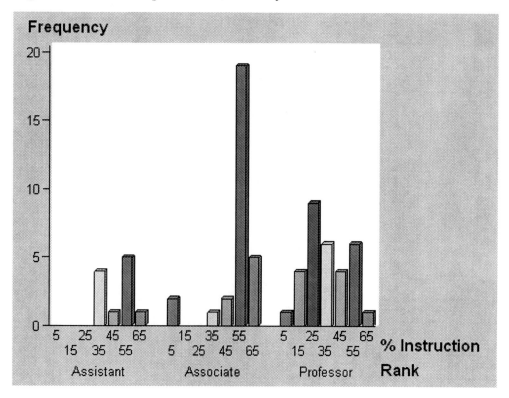

We next add code to compute the kernel density for professional activity and for service. First, however, we need to modify the names of the variables.

Display 2.18. Modify Variable Names

Professional_activity	Service
44	6
60	10
55	7
25	23
41	8
56	9
36	17
36	14
22	12
29	15
59	6
38	12

```
Proc kde data=sasuser.sortfacultyworkload;

Univar Instruction/gridl=0 gridu=100
out=sasuser.kdeinstructionbyrank;

univar professional_activity/gridl-0 gridu=100
out=sasuser.kderesearchbyrank;

univar service/gridl=0 gridu=100
out=sasuser.kdeservicebyrank;

By rank;

Run;
```

This time, we get three new datasets. We construct line plots in a similar fashion.

Display 2.19. Project Diagram for KDE Code

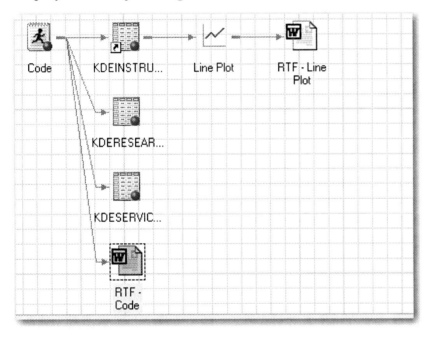

The graph for research shows that associate professors correspondingly have the lowest peak research percentages; assistant professors have the highest.

Figure 2.19. Kernel Density Estimation of Research by Rank

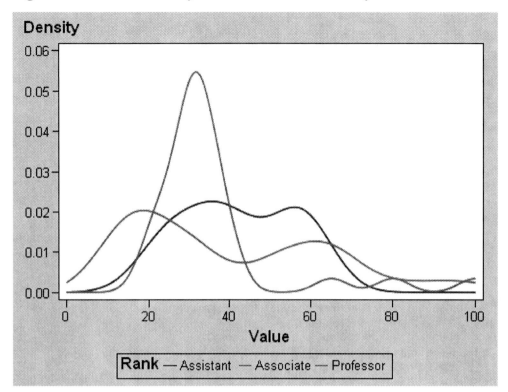

In contract, full professors have the highest levels of service; associate professors have values only slightly higher compared to assistant professors. In addition, full professors have considerable variability in their service assignments.

Figure 2.20. Kernel Density Estimation of Service by Rank

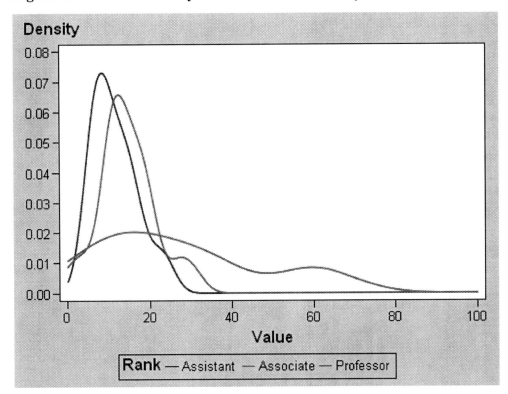

2.4. Exercise

1. Use year to examine faculty workload. First sort by year and then write the code in a code node.
2. Using the student survey data, compute a kernel density of the number of hours claimed for study.
3. Again using the student survey data, compute a kernel density of the number of hours for study by course level.
4. Using your choice of a MEPS database, do kernel density estimates of the payment variables. Then do it by Gender and by Race.

2.5. Appendix

The kernel density estimate is defined by the equation:

$$\hat{f}(x) = \frac{1}{na_n} \sum_{j=1}^{n} K\left(\frac{x - X_j}{a_n}\right)$$

where n is the sample size, K is a known density function, and a_n is a constant depending upon the size of the sample that controls the amount of smoothing in the estimate. Note that for most standard density functions K, where x is far in magnitude from any point X_j, the value of will be very small. Where many data points cluster together, the value of the density function will be high because the sum of $x-X_j$ will be small and the probability defined by the kernel function will be large. However, where there are only scattered points, the value will be small. K can be the standard normal density, or the uniform density. Simulation studies have demonstrated that the value of K has very limited impact on the value of the density estimate. It is the value of the bandwidth, a_n, that has substantial impact on the value of the density estimate. The true value of this bandwidth must be estimated, and there are several methods available to optimize this estimate.

PROC KDE uses only the standard normal density for K but allows for several different methods to estimate the bandwidth, as discussed below. The default for the univariate smoothing is that of Sheather-Jones plug in (SJPI):

$$h = C_3 \left\{ \int f''(x)^2 dx, \int f'''(x)^2 dx \right\} C_4(K) h^{5/7}$$

where C_3 and C_4 are appropriate functionals. The unknown values depending upon the density function f(x) are estimated with bandwidths chosen by reference to a parametric family such as the Gaussian as provided in Silverman:

$$\int f''(x)^2 dx = \sigma^{-5} \int \phi''(x)^2 dx \approx 0.212 \sigma^{-5}$$

However, the procedure uses a different estimator, the simple normal reference (SNR), as the default for the bivariate estimator:

$$h = \hat{\sigma}\left[4/(3n)\right]^{1/5}$$

along with Silverman's rule of thumb (SROT):

$$h = 0.9\min[\hat{\sigma},(Q_1 - Q_3)/1.34]n^{-1/5}$$

and the over-smoothed method (OS):

$$h = 3\hat{\sigma}\left[1/70\sqrt{\pi}\,n\right]^{1/5}$$

Chapter 3. Theory of Hypothesis Testing

3.1. Introduction 95

3.2. Hypothesis Parameters 97

3. Basics of Hypothesis Testing 107

3.4. Example of Faculty Workload 112

3.6. Sample Problem 125

3.7. Exercises 129

3.1. Introduction

In this chapter, we give the basic information on inference, and how to perform inference using SAS in Enterprise Guide. We examine chi-square goodness of fit, and basic t-tests. We also consider one-way analysis of variance. We start by defining a null and alternate hypothesis. These two hypotheses split the universe. That is, the two hypotheses are mutually exclusive and every element in the universe must be contained within one of the two hypotheses. Prior to examining some medical examples, we will first consider some common instances of hypothesis development.

Example of Hypothesis Definition:

Any jury trial requires a decision to be made between two alternatives:

$$H_0: \text{defendant is innocent}$$
$$H_1: \text{defendant is guilty.} \quad (1)$$

However, it could just as easily be written as

$$H_0: \text{defendant is guilty}$$
$$H_1: \text{defendant is innocent.} \quad (2)$$

In fact, (1) is the tradition of English Common Law and is the means by which justice is determined in the American system. On the other hand, (2) is the tradition of the Napoleonic Code and is used throughout much of Europe.

Because of the nature of hypothesis testing, H_0 is always assumed true until the juror must abandon it in the face of overwhelming evidence. Thus in the American system, a defendant is always "presumed innocent." Just as obviously in the Napoleonic Code, a defendant is "presumed guilty." The standard of proof can also differ. For the American civil trial, a "preponderance of evidence is required." For a criminal trial, "evidence beyond a reasonable doubt" is required. It is up to each individual juror to determine whether the standard threshold is achieved so that H_0 can be rejected.

Note that if H_0 is not rejected then a defendant is declared "not guilty." In the American system, a defendant is not called upon to prove his innocence. Thus if not rejected, the null hypothesis is not accepted and the defendant is not proclaimed innocent.

The purpose of stating the hypothesis in this fashion is to minimize the Type I error, or the probability of being wrong if H_0 is rejected. In (1), the error is small that an innocent person will be convicted. However, it is still a possibility and cannot be completely removed. However, the cost is such that many guilty persons will go free. In the Napoleonic Code, the error is very small that a guilty person will go free. However, many of the innocent will be convicted.

However, in any case, there is always the possibility of a "hung jury" where no decision can be reached. A mistrial is declared and a decision is made whether to replicate the trial in the hopes of reaching a conclusion, or to dismiss on the basis that a conclusion probably will never be reached.

3.2. Hypothesis Parameters

Before the data for any test are collected, there are four different parameters which must be considered:

> Type I or α error
>
> Type II or β error
>
> Effect size
>
> Sample size

The four variables are closely related so that three are specified by the investigator and the fourth is computed as a function of the other three.

Type I error

The purpose of any statistical hypothesis test is to make an inference about a population by using a relatively small sample. The possibility of error always exists. Since it cannot be eliminated, it should be controlled. Type I error is defined to be the probability of being wrong if the investigator rejects H_0. Typically, $\alpha = .05$ is considered significant and $\alpha = .01$ is considered highly significant. However, there is more flexibility built into the system that can be used. For example, suppose a treatment is under consideration for patients in the terminal stages of a disease. Then the hypothesis test is stated

> H_0: treatment is not effective
>
> H_1: treatment is effective.

Suppose the p-value is 0.10. This is not usually considered high enough to reject H_0. However, should treatment be denied because the "rule of thumb" value of .05 has not been attained? For terminal patients, these strict guidelines should probably be relaxed.

Consider a hypothesis test that has only a simple alternative:

> H_0: unborn child is a girl
>
> H_1: unborn child is a boy.

Suppose further that the test statistic will be the heartbeat as determined by an ultrasound stethoscope. It is known that the infant heartbeat of a boy is generally lower than that of a girl (Figure 3.1).

Figure 3.1. Hypothesis Test for Sex of Infant

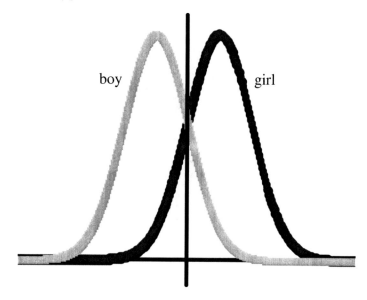

Then the Type I error of $\alpha = .05$ is shown by Figure 3.2.

Figure 3.2. Type I Error

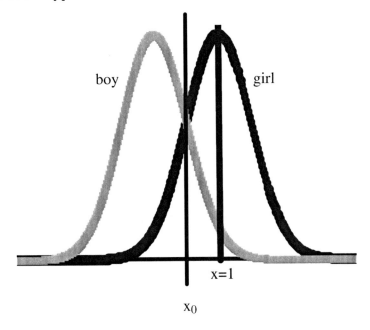

Therefore, if the heartbeat falls anywhere below x_0, it can be concluded that the child is not a girl (rejecting H_0) with $\alpha < .05$ error.

However, suppose that the heartbeat falls somewhere to the right of the value x_1. Then it can be concluded that the child is a girl. In other words, the null hypothesis can be accepted with probability of error β. The value of β represents what is called the Type II error (Figure 3.3).

Figure 3.3. Type II Error

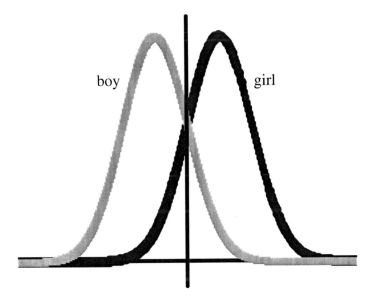

Type II error

More importantly, type II error should also be considered, although it is usually ignored. Type II error is defined to be the probability of being wrong if the investigator accepts H_0. Unfortunately, if the α error is low, the β error can be quite high. Therefore, considering the question posed above for a medical treatment:

H_0: treatment is not effective

H_1: treatment is effective

if the value of α is increased to 0.10, then the probability of denying an effective treatment to a terminal patient decreases. If the value of β is not computed then it is not possible to decide whether to accept H_0. Therefore, the best that can be decided is whether to reject H_0 or to fail to reject H_0. Failing to reject H_0 is often assumed equivalent to accepting H_0, although it is not. In fact, independent of the size of the sample, the maximum possible β error is always $1 - \alpha$. Thus if $\alpha = .05$, it is possible to have a β error of 0.95. Only by defining an effect size can α and β be simultaneously reduced.

In the statistical terminology, the power of the test is defined to be $1 - \beta$. Typically, in a medical study, α is considered acceptable at 0.05 and β is acceptable at a level of 0.20. In this case, the power of the test is stated as 80%.

Consider the hypothesis

H_0: child is a girl

H_1: child is a boy.

Only in the extreme situations depicted can a choice be made between H_0 and H_1; for any value between x_0 and x_1, the hypothesis test is undecidable, even if we know that either H_0 or H_1 must occur.

Consider now a more complex hypothesis. Suppose the test is to determine whether a population has an average blood pressure of 120. Then the hypothesis test is

H_0: u = 120

H_1: u ≠ 120

In this case, the null hypothesis is hard to see amongst the alternative choices for H_1 (Figure 3.4).

Figure 3.4. General Hypothesis Test

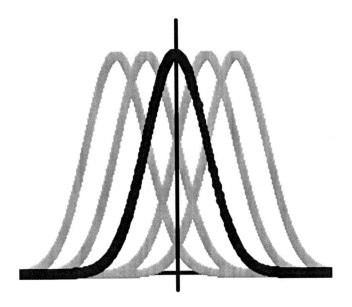

The purpose of sampling is to choose where the sample mean belongs. Suppose it occurs at an extreme (Figure 3.5).

Figure 3.5. Mean at Extreme Observation

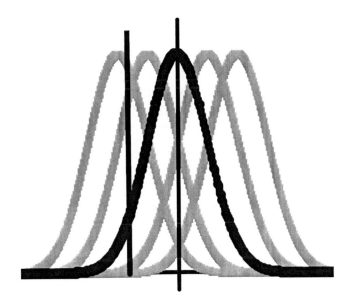

Then not only is it extremely improbable for an extreme observation to belong to a population with mean 120, it is highly probably that it belongs to a population with some other mean. Therefore, in making a choice, you go with the odds. However, if the sample mean occurs closer to the value of 120, then it becomes undecidable. (Figure 3.6).

Figure 3.6. Mean Near H_0 Value

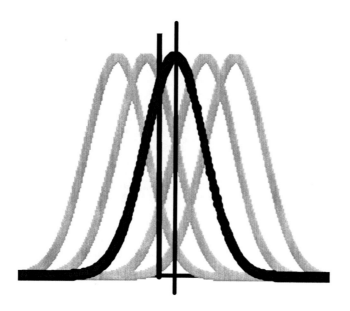

There is too much overlap with the distribution centered at 120 to say that it does not belong but neither can it be said that it does. In fact, since $H_1: u \neq 120$ includes everything but the distribution centered at 120, there is no way to have sufficient evidence to reject it.

Regardless of the sample size, there is a high probability of Type II error independent of the choice of n. There is only one way to eliminate this problem and that is to create a boundary c so that $|u_0 - c| < 120$ for all values of u_0. This boundary is called the effect size.

Effect Size

If a new treatment increases patient survival by 1 day, does it provide sufficient improvement to be worthwhile? Statistical significance cannot determine whether the size of the difference is important; that question must be answered by the investigator. The effect size is defined to be the point at which the difference between treatments becomes important. Once the effect size is defined, the hypothesis can be rewritten (when limited to two treatments)

$$H_0: u_1 = u_2$$

$$H_1: u_1 - u_2 > u' \quad \text{or} \quad u_2 - u_1 > u'$$

where u' is the effect size. Once a specified effect size is given, a sample size can be computed so that both the a and b errors remain small.

Consider again the hypothesis test

$$H_0: u = 120$$
$$H_1: u \neq 120,$$

the value c can be represented on the graph by Figure 3.7.

Figure 3.7. Definition of Effect Size

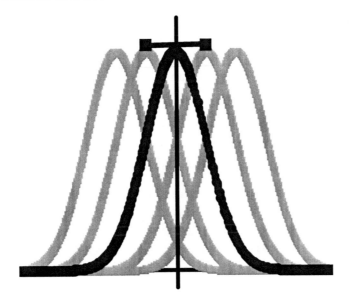

Then and only then is it possible to separate the distributions for H_0 from the distribution for H_1. In this case, the test becomes

$$H_0: u = 120$$
$$H_1: u < 120-c \text{ or } u > 120+c.$$

This is the "good enough principle." Since the larger the sample, the smaller the variance, a sample can be chosen sufficiently large so that the distributions look like that in Figure 3.8.

Figure 3.8. Sample Size Given Effect Size to Isolate H_0.

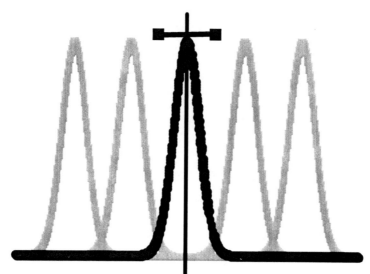

In this case, regardless of where the mean occurs, it can be identified as belonging to one of the distributions.

The decidability of the hypothesis test depends upon the choice of c. It is chosen as an "interval of indifference." In other words, even if the true mean is equal to 120.5, the result is so close to 120 as to have a meaningless difference. This value is called the effect size.

At the same time, it is possible to have a sample size so large that the effect size becomes virtually zero. In this case, any null hypothesis will be rejected simply because of the large size of the sample. Although it is usually the case that a study suffers from a sample size that is too small, Meehl [1990] describes a study so large that any variable discussed was statistically significant. Therefore, the effect size should always be stated and considered. The sample size should always be computed with regard to this effect size.

Sample Size

The smaller the effect size, the larger the needed sample size to keep both α and β under control. If the effect size is too small, it is possible that the required sample size is too large for an experiment to be conducted except for prohibitive cost. On the other hand, if the sample size is limited by cost then the effect size may be so large that H_0 can only be rejected in such extreme situations that it would be virtually obvious without any statistical analysis.

Therefore, effect size is usually identified as a percentage of the population variability with the corresponding sample size computed for small, medium, and large effect sizes. Once this has been done, the null hypothesis can either be accepted or rejected depending upon the outcome of the collected data.

If the effect size is not considered prior to the start of an experiment, it is often the case that the effect size is so large as to render the experiment meaningless. For example, in a study of 71 randomized trials it was found that 50 of them had less than 90% power of detecting a 50% improvement. Thus, if a surgical procedure were to result in the survival of 50% more of the patients, the hypothesis test would not detect this as statistically significant.

The Problem of Generalizations

Hypothesis tests tend to demonstrate differences in averages between populations. The parameters of α, β, n, and ES for ANOVA assume that the populations under study are relatively homogeneous. Generally, by examining enough variables, this homogeneity will be valid. However, if any one segment of the general population is omitted from the study, then it is nearly impossible to generalize any results to that missing segment of the population. As is well known, different subgroups of different racial and national origins have different susceptibilities to disease and do respond differently to treatments.

This has absolutely been true concerning treatment of the heart. In any given sample, the occurrence of heart attacks remains rare. To study the efficacy of any treatment requires a longitudinal study. Because of cost considerations, there is always an interest in reducing both the length of the study as well as the size of the sample. Yet the sample size must be large enough to have a detectable effect size. The solution to the cost problem is to study extremes within the population. Thus the majority of studies on the heart deal only with segments of the population labeled high-risk with generalizations made to the rest of the population. Since pre-menopausal women have relatively low risk, they are usually not part of the sampled population.

The use of mongrel dogs in animal research makes the practice more realistic in making comparisons to the human population. However, as heart disease deals with an extreme in the population, it is also useful to investigate individual dogs with extreme outcomes. Averages can never be used to predict the outcome for any one individual.

To deal with the problem of generalizations, it is important first to determine if the stated hypothesis test will actually satisfy the objective of the study. It is also important to determine if the sample population selected matches the population target under study. These two things must be done prior to any study subjects being chosen. The results must be fully incorporated into the study design. A specific hypothesis will not confirm a vaguely stated theory regardless of the computer p-values.

Another thing that must be considered is the difference between statistical significance and practical significance. The purpose of performing a hypothesis test is to advance knowledge and to provide direction in research. If the experiment does neither of these things then it is better left undone. It is also wise to keep in mind the limitations of hypothesis testing. It can answer the question of what happens but can contribute very little to the questions of why or how.

Central Limit Theorem

Most of hypothesis testing is based upon the Central Limit Theorem. Consider, for example, a time when your instructor indicates that he/she graded 'on the curve'. Sometimes, your instructor drew the bell curve, and indicated just where the grades would fall (Figure 3.9).

Figure 3.9. Normal Distribution

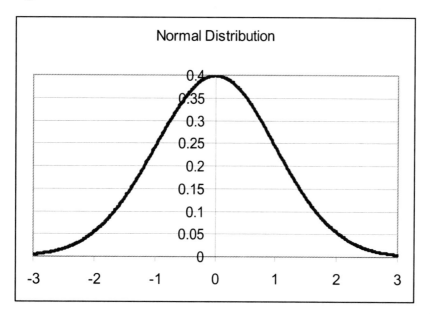

That was, in fact, an incorrect use of the bell-shaped curve, unless your instructor had a good idea that the class was relatively homogeneous.

The central limit theorem is concerned with a course with multiple sections. Consider the average grade for each section. If all of the averages are plotted on a graph, the result would look bell-shaped regardless of whether any individual class resembled a bell-shaped curve.

In other words, the Central Limit Theorem is concerned with the "average of averages". If we collect several data points, and compute the average of those data points and then duplicate this process so that we have a collection of averages, the graph of these averages will look bell-shaped.

The Central Limit Theorem is important because it is the basis of all of the inferential statistics we will be studying in this course. Every time we make an inference, we make the assumption of a bell-shaped (or normal curve) in order for the test of our inference to be valid.

3. Basics of Hypothesis Testing

The simplest hypothesis test to consider is whether the mean of a population is equal to a particular value,

H_0: $u = u_0$

H_1: $u > u_0$ or $u < u_0$ or $u \neq u_0$

However, it rarely comes up in practice and we will not be discussing it in great detail. The next hypothesis test compares the means of two populations to each other:

H_0: $u_1 = u_2$

H_1: $u_1 \neq u_2$

This hypothesis is a special case of comparing two or more population means:

H_0: $u_1 = u_2 = u_3$

H_1: at least two of u_1, u_2, u_3 are not equal

Analysis of variance is a statistical method for comparing multiple populations while simultaneously considering a number of population factors that may have impact on the final results. A typical research question is whether or not a particular treatment provides significantly better results than do other treatments. Such a question can be translated into a statistical hypothesis test of the form:

H_0: $u_1 = u_2 = \ldots = u_k$

H_1: at least two treatments are different

Note that the hypothesis test deals only with a comparison of the average treatment effects across populations and that the alternate hypothesis only requires that at least two of the k treatments differ. Thus the general ANOVA test examines the general situation. Only if H_0 is rejected should the data be examined more closely to determine just where differences occur. In this case, there are a variety of multiple comparison tests that can be used. There are so many because not one of them is better than any of the others in all situations. The investigator must decide what is of most importance. There are three main choices:

- Find as many differences as possible.

Cost: Greatly increase the cumulative Type I error as the number of comparisons increases, and differences will be declared where none exist

- Keep the Type I error under control

Cost: May find no differences even where they exist.

- Also keep the Type II error minimized

Cost: May find no differences even where they exist.

The best strategy is to use several multiple comparisons and then make a judgment based upon the extent of optimizing outcome and minimizing cost.

If the sample selected is totally random, then the structure of a one-way analysis of variance as described above is sufficient. However, in medical studies, the sample is rarely random. Therefore, it becomes important to identify population characteristics that can contribute to the final outcome. If the sample is relatively small and the subjects are different, this could result in considering each subject as a separate class in the analysis of variance.

For example, consider a recent study conducted to determine differences in blood flow in different blood vessels when these different vessels were opened and closed. The study was conducted on dogs. The desired outcome was to compare four different flow mechanisms:

$$LAD$$
$$LAD \text{ vs } SVG$$
$$LAD \text{ vs } PIMA$$
$$LAD \text{ vs } DIMA$$

The null and alternate hypotheses are written as follows:

$$H_0: u_{LAD} = u_{LAD \text{ vs } SVG} = u_{LAD \text{ vs } PIMA} = u_{LAD \text{ vs } DIMA}$$

$$H_1: \text{at least two means are different}$$

There were three additional variables considered that might contribute to the outcome:

Choice of dog (12 dogs were used; all were mongrels)

Type of cycle (systolic, diastolic, and whole cycle were measured)

Method of measurement (dogs 1-7 were measured by hand; dogs 8-12 by computer).

In addition to the type of dog used, the part of the cycle where measurements were taken could have an effect on the final outcome. The need to examine the method of measurement was necessitated by a change made in the middle of the experiment.

Ideally, the choice of variables to be studied should be made prior to the start of an experiment and should be discussed in consultation with a statistician. The number of variables studied can have an impact on the computation of effect size and sample size, which should also be computed before the start of any experiment.

First, a general ANOVA table is given to determine whether any of the variables were significant. The SAS software used provided the following summary:

Source	DF	SS	MS	F Value	p-value
Model	16	150657.46	9416.09	119.92	0.0001
Error	415	32585.5	78.52		
Total	431	183242.96			

R-Square: 0.822 Coefficient of Variation 36.94

Most of the information should be included in a written result. The p-value of 0.0001 indicates an effect not occurring as a result of chance. However, the R-Square value and the coefficient of variation provide more valuable information. Roughly, the R-Square value indicates what percentage of the variability in the outcome variable can be attributed to the variability of the input variables. 82% is a fairly high value indicating a strong relationship. The coefficient of variation is the ratio of the mean to the variance of the outcome variable. The second summary table provided by SAS gives an indication of the effect of each individual variable:

Source	DF	SS	MS	F Value	p-value
Procedure	3	12682.53	4227.51	53.84	0.0001
Method	1	50663.51	50663.51	645.24	0.0001
Dog(Method)	10	79204.91	7920.49	100.87	0.0001
Cycle Type	2	8106.52	4053.26	51.62	0.0001

Note that the first column is labeled "DF" (degrees of freedom). This gives an indication of the number of classes in each variable. There were k = 3 + 1 procedures, m = 1 + 1 methods, and c = 2 + 1 different cycle types collected. The value for dog (method) = # dogs (12) – # methods (2) = 10. The sum total of 3 + 1 + 10 + 2 = 16 is the value given in the first table when the variables were considered as a collective whole.

Since the dogs were mongrels, they were considered individually because mongrels can vary substantially from each other. Because dogs 1–7 were measured by hand and dogs 8–12 by computer, there is no way to compare the two methods on the same dog. In statistical terms, the value of dog is nested within the value for method.

The ANOVA procedure computes nested effects differently from standard effects. It is possible that there could be interactive effects as well. That is, procedure 1 could have a greater effect than the others for the systolic part of the cycle, but procedure 2 could have a greater effect for the diastolic part of it. Such interactive effects can be investigated, but such an investigation requires a larger sample size. Because only 12 dogs were used in this study, such interactive effects were not computed. However, it should be decided beforehand whether such interactions should be examined so that the sample size can be adjusted accordingly.

Since all variables are significant, it becomes important to separate the effect from the procedure. In this case, the multiple comparison tests are no longer valid. They can only be used with one specified variable. Statistically, it can be done but the cumulative Type I error increases substantially with the number of comparisons made. Therefore, it is best to keep the number of comparisons to a minimum and to decide beforehand which comparisons are to be made.

In this study, the four procedures were compared in terms of flow. The adjusted means were equal to

LAD	32.75
LAD vs SVG	19.59
LAD vs PIMA	24.24
LAD vs DIMA	19.38

At a = .05 (cumulative error of α = .18), LAD is different from the other three procedures and PIMA is different from the other two.

3.4. Example of Faculty Workload

Up to this point, we have just investigated the faculty workload data using data summaries and graphics. We now want to test the hypothesis:

H_0: $u_{assistant} = u_{associate} = u_{full\ professor}$

H_1: at least two are different

In order to make this test, we use the general linear model available in Enterprise Guide (Display 3.1).

Display 3.1. Menu for Linear Models

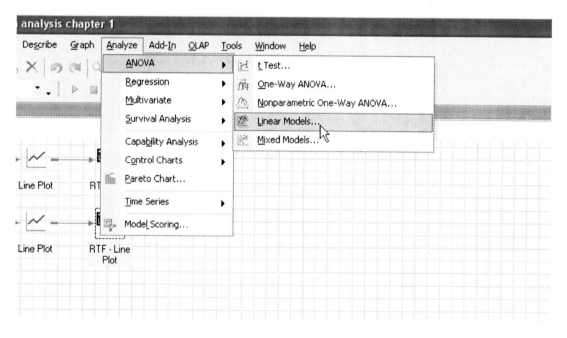

We want to compare the proportion of instruction by faculty rank (Display 3.2).

Display 3.2. Screen for Linear Models

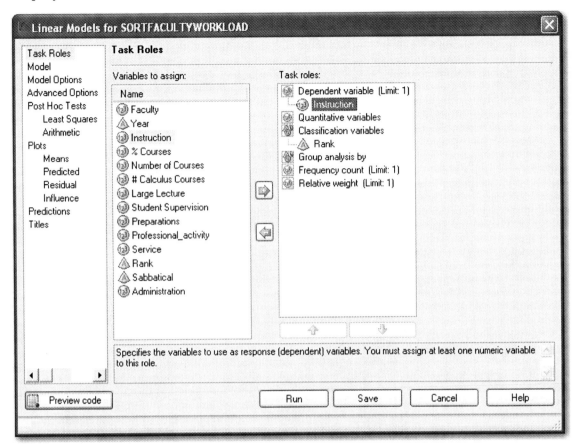

Note that rank is listed as a classification variable. Nominal variables are always defined as classification; interval variables are always defined as quantitative. Ordinal variables can be defined in either location. To see the difference in results, it is suggested that you use ordinal variables in both locations and see what happens. We next need to define the model (Display 3.3).

Display 3.3. Definition of Model

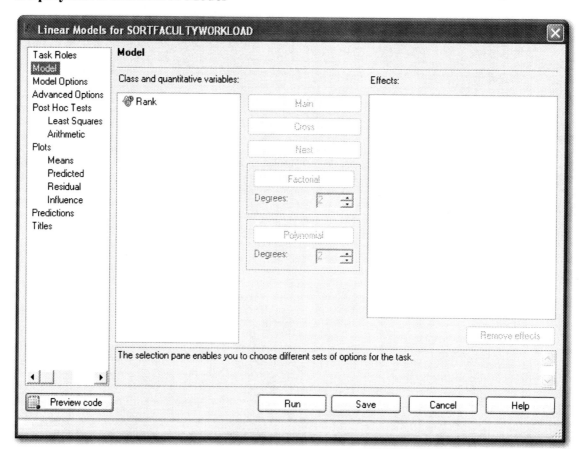

We need to move rank over to the right hand side as a main effect. We do this by highlighting rank and then pressing the button, **MAIN** (Display 3.4).

Display 3.4. Moving Rank to Effects

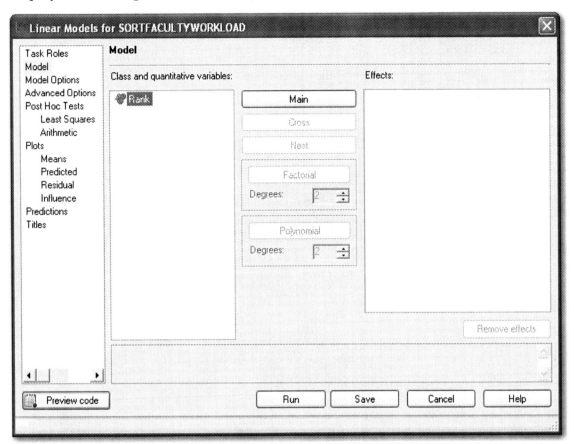

We then want to move to the Arithmetic Post Hoc Tests (Display 3.5).

Display 3.5. Arithmetic Tests

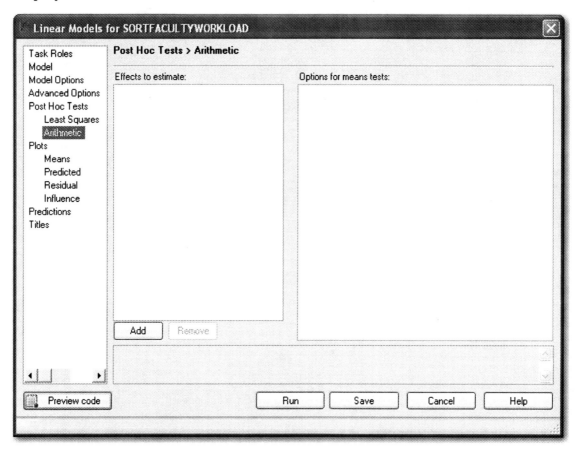

By hitting the 'Add' button, a number of options appear (Display 3.6).

Display 3.6. Defining Tests

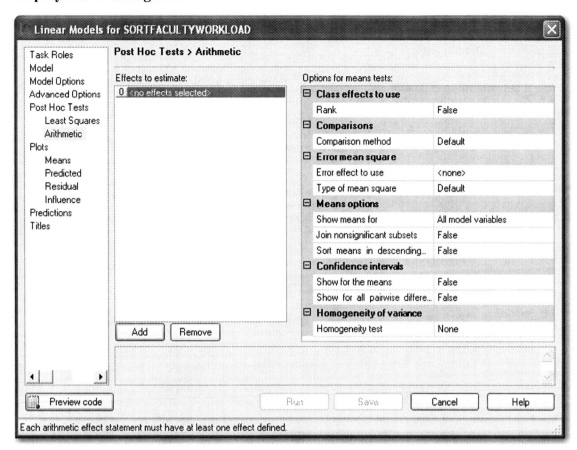

The options we want to use are shown in Display 3.7.

Display 3.7. Defining Options for Arithmetic Tests

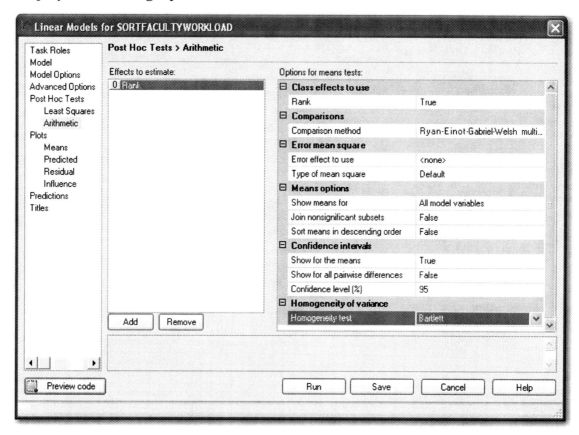

Note that there is a test for homogeneity of variance. This test assumes that the variability among the three faculty ranks is the same. This assumption may or may not be valid, and we need to test it.

We also want to look at some plots. There are additional assumptions that need to be considered, and we will discuss them at length shortly.

Display 3.8. Plots Available for Checking Model Assumptions

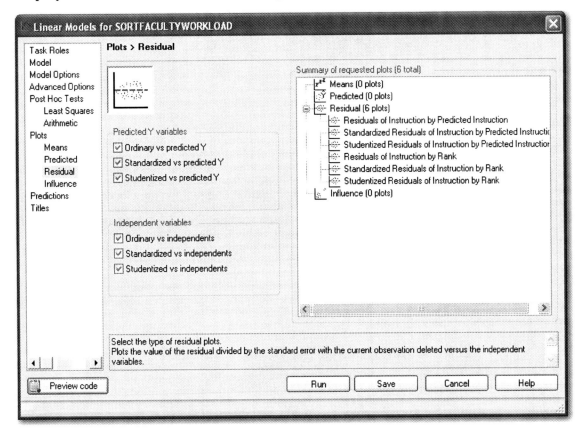

Display 3.9. Project diagram for Linear Models

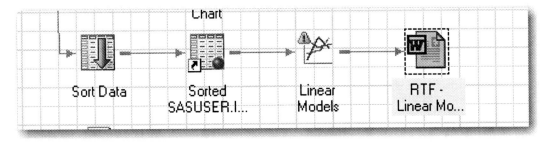

We get the following results (Tables 3.1-3.8).

Table 3.1. Summary of Class Variables

Class Level Information		
Class	Levels	Values
Rank	3	Assistant Associate Professor

The first section gives the class variables, and the number of levels for that variable.

Table 3.2. Number of Observations Used

Number of Observations Read	71
Number of Observations Used	71

We next get the number of the observations in the dataset, and the number of observations used in the analysis. If an observation has missing values, it will not be used.

Table 3.3. Overall Summary of the Model

Source	DF	Sum of Squares	Mean Square	F Value	Pr > F
Model	2	4591.49349	2295.74675	10.10	0.0001
Error	68	15449.46425	227.19800		
Corrected Total	70	20040.95775			

We next get a summary of the model. The p-value of 0.0001 suggests that there is a difference in instructional proportions by faculty rank.

Table 3.4. R-Square and Coefficient of Variation

R-Square	Coeff Var	Root MSE	Instruction Mean
0.229105	36.22848	15.07309	41.60563

The R-Square suggests that 23% of the variability in instruction is explained by the difference in faculty rank. However, that means that 77% of the variability in instruction is still not accounted for in the model.

Table 3.5. Type I Sum of Squares

Source	DF	Type I SS	Mean Square	F Value	Pr > F
Rank	2	4591.493493	2295.746747	10.10	0.0001

The next table gives a clarification and breakdown by variable. Since there is only one variable in the model, this table duplicates the information already provided above. Note that there is a table for type I SS and Type III SS. These tables will be identical if there is only one variable. Therefore, we will discuss the differences at a later time.

Table 3.6. Type III Sum of Squares

Source	DF	Type III SS	Mean Square	F Value	Pr > F
Rank	2	4591.493493	2295.746747	10.10	0.0001

The next table provides estimates of the coefficients used in the linear equation defined by Instructional percent=32.55+13.81 for assistants and 32.55+16.93 for associates and 32.55+0 for full professors. The P>|t| gives the likelihood that the coefficient value is zero. Here, it is not likely.

Table 3.7. Parameter Estimates

Parameter	Estimate		Standard Error	t Value	Pr > |t|
Intercept	32.54838710	B	2.70720665	12.02	<.0001
Rank Assistant	13.81524927	B	5.28992739	2.61	0.0111
Rank Associate	16.93437152	B	3.89401872	4.35	<.0001
Rank Professor	0.00000000	B	.	.	.

Note: The X'X matrix has been found to be singular, and a generalized inverse was used to solve the normal equations. Terms whose estimates are followed by the letter 'B' are not uniquely estimable.

Table 3.8. Test for Equal Variances

Bartlett's Test for Homogeneity of Instruction Variance			
Source	DF	Chi-Square	Pr > ChiSq
Rank	2	2.0352	0.3615

We have a potential problem here. When the matrix of the model is singular, there can be infinitely many solutions. The question is whether we can eliminate this singularity by slightly altering the model, or whether we have to acknowledge it and see if it has impact on the model outcomes. The test for homogeneity of variance has a large p-value of 36%. Therefore, we can safely assume that this assumption is valid.

Figure 3.9. Residuals of Actual Instruction by Predicted Instruction

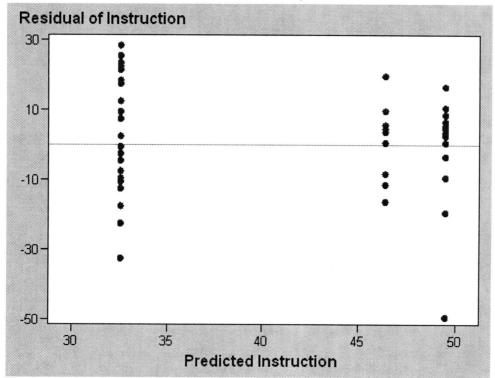

The residuals should show a random pattern. Since the height of the line does not increase with the value of predicted instruction, the assumption of randomness is valid.

Figure 3.10. Standardized Residuals of Instruction

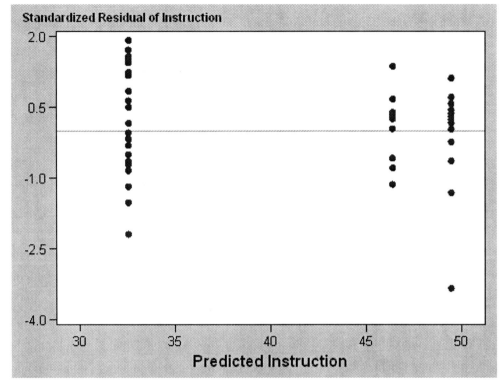

The standardized residuals show the same pattern, just changing the scale on the y-axis, with the value of 0.0 at the center.

Figure 3.11. Residual of Instruction by Rank

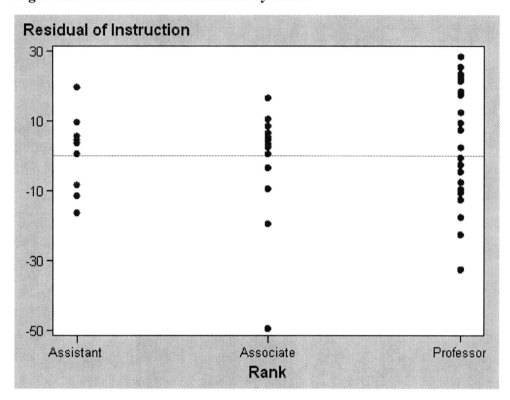

Figure 3.12. Standardized Residual of Instruction by Rank

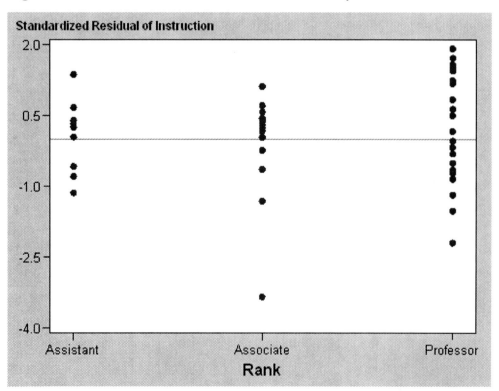

3.6. Sample Problem

An Analysis of Average Monthly Sales for PPCO

Abstract:

This paper will analyze the monthly sales for PPCO spanning from the year 1994 until 2003 in order to see if there is any correlation in the sales on a monthly basis. It will compare and contrast the sales and see how the sales varied as the years passed by. This will show if there are truly busy and slow months for the business. Conclusions will be drawn as to why these sales differ from month to month and year to year.

Introduction:

This paper will analyze the dataset dealing with the monthly sales for PPCO from 1994 till 2003 in order to show correlations in the overall sales and discover patterns if any exist. After running several tests with the generalized linear model, it doesn't seem that any of the months have a strong relationship due to all months returning p-values greater than .4. By examining the data using multiple line graphs for visualized study, it is clear that the months do follow a trend in their overall sales. The months will be examined as a whole and then divided appropriately into groups by which months seem to follow the same trend.

Method:

The dataset consists of the total earnings for PPCO spanning from 1994 till 2003. Most of the data from the last 3-4 years was easily accessible by looking into the library of the previous Commentaries. The commentary is a company newsletter that is distributed every month to the company's employees. In each issue, the company president will give a short message that details specifics of plant production and also a comparison of sales over the last month compared to the same month in the previous year. After searching through the archives, it was discovered that Mr. X actually started giving this particular detail in the July, 2001 issue. Therefore, the remainder of the data were collected by interviewing one of the company accountants.

These data will be analyzed with the help of Enterprise Guide. A majority of the information given in this paper will be from summary statistics and from data visualization techniques. The generalized linear model was used to look for strong correlations between the monthly sales since this data set deals with interval data. The general linear model returned data that did not reflect a strong correlation between the months with p-values greater than .4. From the summary, it is seen that there were major increases in sales over the years. In early 1994, the company averages approximately 9.4 million as opposed to closing out 2003, making around 14 million. The company seemed to reach its pinnacle in the summer and fall of 2000 with monthly earnings in the 17 million range. Then the company dropped in earnings to average between 13 and 15 million for the next 3 years. One of the reasons for the slow down in profit could be attested to the terrorist attacks in September of 2001. This seems to have caused a halt to the steadily increasing sales the company was experiencing.

Results:

While studying the data, it seemed an important task to find trends, if any existed. In order to perform this task, a test was run against the data set by using the general linear model. It revealed that a strong correlation between the monthly sales does not exist due to a p-value returned greater than 0.4. The only months that seemed remotely related were January and February.

Due to the results of the general linear model, it was necessary to examine these data from a visualization standpoint. In order to do this, a graph was produced to examine all the months compared together. This can be seen in Figure 1. It is apparent that some months sales vary greatly from other months. There is a very distinct peak in sales for one particular year: 2000. This seemed to be the pinnacle year for PPCO. Their sales reached an all time high of over 17 million dollars. No other year can quite compare to that year.

Figure 1. Sales by Month

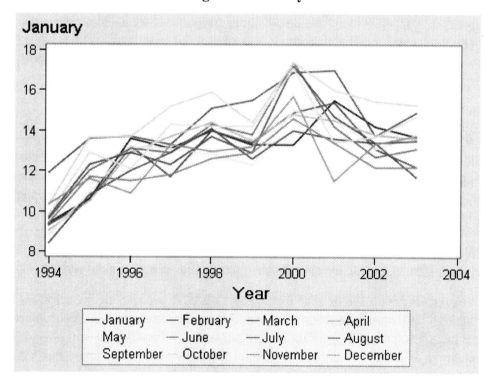

As discussed before, it is most likely that the cause for the downslide in sales resulted from the terrorist attacks on the United States in September, 2001. This affected a lot of businesses; some good, some bad. It was overall bad for the advertisement industry. Advertisements were down, which affected the page counts in magazines. When the page counts went down, so did profits due to the magazines not costing the usual amount to print.

In order to better view and analyze the data, it was necessary to divide it into groups. They are divided by the months that have the strongest correlation to each other, or the sales that were mostly consistent over the 10 year period. They were easily divided into two groups. Both groups are displayed in the Figure 2 and Figure 3.

Figure 2 displays the months that generally have the least sales. These months are January, February, April, May, July, November and December. As is seen here, January and February are listed under the same category not only because it is apparent from the data visualization, but also from the results of the general linear model. It could be easily explained why January, February, November, and December are on this list. Since they are winter months, sales could be normally down at those times of year. People may not want to read magazines as much. It doesn't seem likely that this would be the case. It would seem that most people would rather read in the winter than in the summer. The days are shorter and are cold in the winter, which would make people want to stay inside and read. There are numerous reasons why sales should be greater in the winter, but the analysis shows otherwise.

Figure 2. Sales in Lower Sales Months

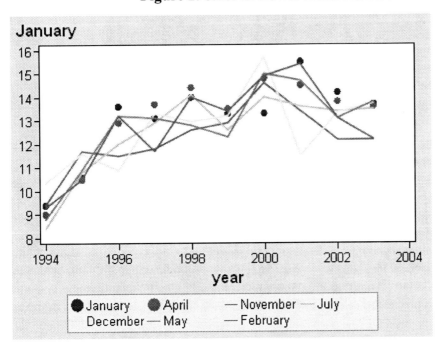

As is expected, Figure 3 hosts those months from the dataset that managed to pull in the most sales over the 10 year span. These months are March, June, August, September and October. The summer and fall months hold a staggering lead in sales over the winter months. Every month listed above was making over 12 million per year since 1997. The business seemed to take on a booming increase since its dismal earnings averaging 10 million in 1994. By today's standards, PPCO's earnings in 1994 would not be enough to stay in business.

Figure 3. Sales in High Sales Months

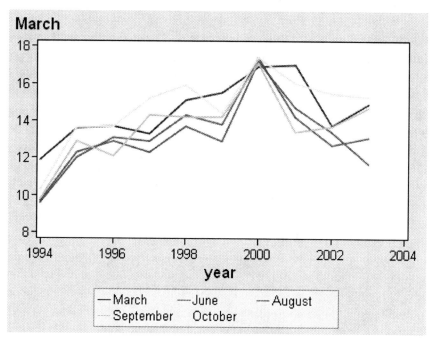

As the sales are examined closely, it is easy to see the large amount of sales procured in 2000. The only plausible reason for the increased sales would be due to the new millennium. There was a big movement over the change of the century that caused many people to take advantage of the situation. It is wise to assume that advertisements and page counts increased greatly during this time frame. The following year, sales took a nose dive and ended up averaging about the same as they did in 1999.

The most helpful tool used in this analysis was the data visualization and the summary statistics portions of Enterprise Guide. The general linear model did help with suggesting that January and February sales were slightly related, but it did not help with determining trends in the remaining months.

Conclusion:

Being that the company is family owned and operated, it is quite surprising to some how successful their business is. With annual sales reaching as high as 188 million over this 10 year span, it is quite amazing to see how successful they are with a seemingly small company. The company struggles hard to keep in business and it is growing every day. With the purchase of new equipment and technology, the company is a very strong competitor in the printing industry. It is very clear that this is a profitable business. As a follow up, it would be a great task to follow this company and see how their sales increase or decrease over the next 10 years. After studying this dataset, it is a good bet that PPCO will continue to be a profitable business.

3.7. Exercises

1. Use the student survey data and define several hypotheses from the dataset. Determine whether the variables you are testing are interval, ordinal, or nominal.

2. Use the MEPS dataset and define some hypotheses. Determine whether the variables you are testing are interval, ordinal, or nominal.

Chapter 4. Types of Statistical Analysis

4.1. Introduction 132

4.2. Data Mining 134

4.3. Cost-Effectiveness Analysis 135

4.4. Analysis of Variance 136

4.5. Processing Quality-of-Life Information 138

4.6. Examples 142

4.7. Exercises 157

4.1. Introduction

The objective generally leads to a specific type of statistical analysis. It is important to have a clear idea of the type of analysis to be used once the data are collected. The method of analysis has a strong influence on the size of the data sample needed as well as on the variables needed to satisfy the objective.

Sample size always needs to be specified before any data are collected. If the sample size is chosen correctly, there will be a definite conclusion at the end of the experiment. If the sample is too small and the results are negative, the experiment is undecidable. If the sample is too large, every outcome will be statistically significant and totally unimportant. Therefore, it is important to choose the sample size with care.

Table 4.1. provides a brief list of the most basic statistical methods:

Table 4.1. Summary of Statistical Methods

Method	Objective
Chi-square Analysis	Determines whether a relationship exists between two characteristics that are defined categorically. Should not be used except in extremely simple studies, or to reduce a large parameter list to a more manageable number.
Analysis of Variance, Linear Regression	To compare two or more types of treatment. A t-test is a simplified version of analysis of variance. This is the most commonly used type of statistical analysis. This technique can be used to account for any specified confounding factor. It can be used to find differences between treatments. It can also be used to show that treatments are equivalent (this requires a larger sample).
Survival Analysis	This can compare patient survival between treatments, also accounting for specified confounding factors. Survival does not necessarily mean life versus death. Instead, it is a method to compare time to an event where the event can be survival, recurrence, etc.
Logistic Regression, Discriminant Analysis, Neural Network Analysis	Usually determines whether certain patient characteristics put a patient at higher risk for an adverse outcome. Once risk is established, the methods also determine just how well patient characteristics can predict outcomes. It is often used in conjunction with chi-square analysis (to reduce the number of factors under consideration).
Factor Analysis,	Used to validate patient data forms. Also used to examine the relationship between large variable sets. It is an

Method	Objective
Cluster Analysis	exploratory method and is not used to investigate specific hypotheses. Used frequently in investigations of quality-of-life
Decision Trees	Primarily for cost-effectiveness analysis.

4.2. Data Mining

With large patient databases and no concrete hypothesis test, it is still possible to perform statistical analysis. Data mining is an investigative procedure to find patterns in the data, patterns that can be validated through further examination. It is a process to first generate hypotheses and then to test them.

Table 2. Process of Data Mining

Procedure	Rationale
Developing an understanding of the application domain, the relevant prior knowledge, and the goals of the end user.	What are the goals? What performance criteria are important? Will the final product of the process be for classification, visualization, summarization, prediction? Is understandability an issue? What is the trade-off between simplicity and accuracy?
Creating a target data set, selecting a data set, or focusing on a subset of variables or data samples, on which discovery is to be performed.	This involves consideration of homogeneity of the data, change over time, sampling strategy, etc.
Data cleaning and preprocessing	This involves the removal of noise or outliers, deciding on strategies for handling missing data fields, etc.
Data reduction and transformation	This involves finding useful features to represent the data, depending on the goal of the task; using dimensionality reduction or transformation methods to reduce the effective number of variables under consideration or to find invariant representations for the data; and projecting the data onto spaces in which a solution is likely to be easier to find.
Choosing the data mining task	Deciding whether the goal is classification, regression, clustering, summarization, modeling, etc.
Data mining	The actual analysis, automated and most often done with software
Evaluating output	Determining what is actually knowledge and what is "fool's gold." Knowledge must be filtered from other outputs. This can be done by performing statistical checks, visualization, or expertise.
Incorporating knowledge into the performance system.	Checking and resolving potential conflicts with previously extracted knowledge.

4.3. Cost-Effectiveness Analysis

An analysis of cost usually depends upon expert opinion of risk rather than upon empirical data. Therefore, research projects that use empirical data, when done properly, will be extremely publishable. There is a difference between an examination of actual cost of treatment, and patient billing. Decide early on which cost will be investigated. This is particularly true for patients with chronic problems who come regularly to a clinic setting for disease management. There are direct costs that need to be collected:

Table 4.3. Types of Costs

Type of Cost	Items for Collection
Direct to patient	Direct cost to the patient such as medication, devices, etc.
Surgical Procedures	Cost to patient of time in OR, time in hospital, etc.
Direct costs of time	Personnel time involved. Personnel needed to treat patient, time needed and cost for each unit of time.
Overhead costs	The costs associated with operating a clinic. This can be used as a cost/day divided by the amount of individual patient time.
Indirect costs (or benefits)	Costs associated with reduced capacity. Time lost from work for treatment. Requires collection of quality-of-life information.

Such a study would have to be longitudinal and prospective. It can answer a simple question such as will one medication be cost-effective by reducing healing time even if it costs more for each use? Such a study can also be used to determine whether angioplasty is more (or less) cost-effective than bypass surgery.

4.4. Analysis of Variance

Analysis of variance is a statistical method for comparing multiple populations while simultaneously considering a number of population factors that may have impact on the final results. It is a generalization of the 2-population t-test. If multiple pairwise t-tests are used in a situation where an ANOVA is more appropriate, it will be less likely that statistical significance will be achieved. A typical research question is whether or not a particular treatment provides significantly better results than do other treatments. Such a question can be translated into a statistical hypothesis test of the form:

$$H_0: u_1 = u_2 = \ldots = u_k$$

H_1: at least two treatments are different

Note that the hypothesis test deals only with a comparison of the average treatment effects across populations and that the alternate hypothesis only requires that at least two of the k treatments differ. Thus the general ANOVA test examines the general situation. Only if H_0 is rejected should the data be examined more closely to determine just where differences occur. In this case, there are a variety of multiple comparison tests that can be used. There are so many because not one of them is better than any of the others in all situations. The investigator must decide what is of most importance. There are three main choices:

- Find as many differences as possible.

  Cost: Greatly increase the cumulative Type I error as number of comparisons increases, and differences will be declared where none exist

- Keep the Type I error under control

  Cost: May find no differences even where they exist.

- Also keep the Type II error minimized

  Cost: May find no differences even where they exist.

The best strategy is to use several multiple comparisons and then make a judgment based upon the extent of optimizing outcome and minimizing cost.

If the sample selected is totally random, then the structure of a one-way analysis of variance as described above is sufficient. However, in medical studies, the sample is rarely random. Therefore, it becomes important to identify population characteristics that can contribute to the final outcome. If the sample is relatively small and the subjects are different, this could result in considering each subject as a separate class in the analysis of variance.

4.5. Processing Quality-of-Life Information

Quality-of-Life is difficult to measure since it is so very subjective. The best way to do so is to collect data from the patients themselves on an ongoing basis. Consider the following short, simple baseline form that can be used for a chronic problem such as heart failure:

1. Prior to your diagnosis of heart failure, were you ☐ unemployed ☐ employed ☐ retired?
2. Prior to your diagnosis, what percent of the time were you absent from work (if employed)?

   ☐ 0% ☐ 1-10% ☐ 11-25% ☐ 26-50% ☐ 51-75% ☐ more than 75%
3. Prior to your diagnosis, what percent of the time were you unable to go about your normal daily routine?

   ☐ 0% ☐ 1-10% ☐ 11-25% ☐ 26-50% ☐ 51-75% ☐ more than 75%
4. Prior to diagnosis did you frequently feel (*check all that apply*) ☐ weak ☐ irritable ☐ tired ☐ depressed ☐ forgetful ☐ feverish ☐ anxious ☐ confused ☐ hungry ☐ thirsty ☐ sad ☐ nervous
5. Prior to diagnosis, you frequently had (*check all that apply*) ☐ nausea ☐ constipation ☐ diarrhea ☐ headaches ☐ muscle or joint aches ☐ shortness of breath ☐ swelling
6. Because of the problems checked in #4 and #5 above, ☐ always ☐ frequently ☐ occasionally ☐ rarely ☐ never had problems going about your normal routine.
7. Prior to diagnosis, how would you rate your overall physical condition?

   (*Circle one value 1-10*)

Very Poor   1   2   3   4   5   6   7   8   9   10 **Excellent**

8. Prior to diagnosis, how would you rate your overall quality of life? *(Circle one value 1-10)*

Very Poor   1   2   3   4   5   6   7   8   9   10 **Excellent**

At each subsequent visit, the questions (slightly modified) can be asked to determine what changes have taken place, and whether treatment has improved the quality-of-life. In addition, specific questions related to a particular illness and treatment can be added.

Chi-Square Analysis

There is no question that treatment is having an impact on patient quality-of-life. The QOL parameters were examined at baseline, at 3, 6, and 9-month intervals. To determine the impact of medication, the 7 treatment groups were divided into interferon (treatments 2,3,6) and no medication (treatments 1,4,5,7). The 1-10 scale of physical condition and quality-of-life was condensed into categories: <5.5 and >5.5.

There were no statistically significant differences between treatment groups at baseline with the exception of physical condition and quality-of-life. However, differences were shown at the followup time periods.

Table 4.4 gives the p-values for differences between treatments at 3, 6, and 9 months followup (using chi-square tests).

Table 4.4. Quality of Life Measures Weak

Characteristic	Baseline	3-Months	6-Months	9-Months
Weak	0.101	0.001	0.001	0.011
Irritable	0.824	0.032	0.001	0.001
Tired	0.21	0.001	0.001	0.001
Depressed	0.78	0.9	0.002	0.05
Forgetful	0.142	0.932	0.103	0.002
Feverish	0.199	0.001	0.001	0.011
Anxious	0.252	0.068	0.006	0.978
Confused	0.115	0.521	0.001	0.014
Hungry	0.587	0.044	0.397	0.011
Sad	0.691	0.464	0.001	0.411
Nervous	0.167	0.044	0.034	0.213
Nausea	0.674	0.001	0.506	-----
Constipation	0.704	0.017	0.003	0.39
Diarrhea	0.996	0.044	0.014	0.002
Headache	0.126	0.001	0.016	0.078
Muscle	0.892	0.001	0.026	0.001
Shortness	0.504	0.009	0.249	0.075
Swelling	0.811	0.788	0.385	0.177
Appetite	0.674	0.001	0.036	0.39
Weight Change	0.721	0.001	0.527	0.163

Characteristic	Baseline	3-Month	6-Month	9-Month
Physical Condition	0.027	0.001	0.001	0.391
Quality-of-Life	0.033	0.001	0.006	0.391

As the 9-month values only had 45 observations, the p-values are somewhat questionable and need to be validated when more data become available. It is troubling that there are differences in quality-of-life and physical condition at baseline. However, it must be recalled that treatments 3 and 7 are not randomized. If the analysis is divided between randomized and non-randomized patients, the results are in Table 4.5.

Table 4.5. Quality of Life Modification

Characteristic	Baseline With Interferon	Without Interferon	3-Month With Interferon	3-Month Without Interferon
Physical Condition	0.273	0.322	0.694	0.436
Quality-of-Life	0.001	0.255	0.269	0.233

As a future analysis, it is of interest to determine whether patients on medication improve in overall quality-of-life once the Interferon treatments are completed.

However, the quality parameters do not predict well the use of medication. To examine the issue of predictability, logistic regression is usually performed. In a logistic regression, the area under the receiver operating curve is only equal to 0.77 (Figure 4.1).

Figure 4.1. Receiver Operating Curve to Examine Effectiveness of Model

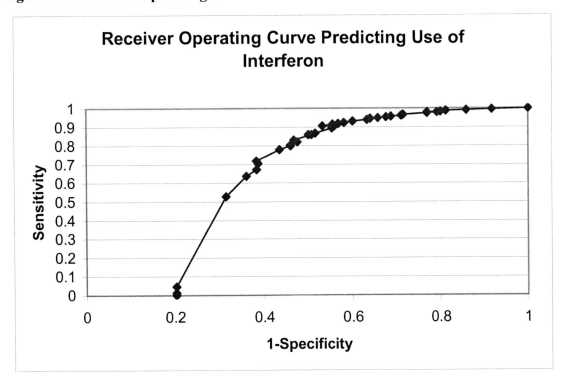

There are other types of statistical methods available that are not taught in this course. It is important to be aware of some of these methods because you will not always be able to work with the data you have using only the methods taught here.

For example, if the outcome variable is the time to an event (the time to acquiring an infection or the time to recurrence of cancer, or the time between a customer getting a cell phone line and changing providers), then the method to be used is survival analysis. If the variable of interest is the monthly phone bill or retail sales by month, or the amount of money to use in an ATM machine given the daily transactions, the method to use is time series analysis. Decision trees are used to classify groups.

4.6. Examples

The first example examines patient compliance for the care of diabetic foot ulcers. The second example examines staff compliance with the treatment of patients with diabetes while undergoing cardiovascular surgery.

1. Introduction

This is a preliminary report of the progress made so far in investigating the data from a foot ulcer clinic. To date, 80 patients from the clinic were systematically sampled to determine distributions of patient diagnosis, compliance, etc. The remaining patients will again be systematically sampled to validate the results from the preliminary sample.

For the foot ulcer clinic, a working knowledge of diabetic foot ulcers was required. It was necessary to convert information from patient charts into a format acceptable to an electronic database. The database underwent several modifications in the early stages, as the material became clearer.

1a. Problems

It should be noted that there were a number of problems dealing with the establishment of the database. Patient record acquisition and translation were among the most daunting and immediate hurdles. For example, it took a week of time just to get a list of patient names. The only electronic list in existence was in the billing files. An attempt was made to narrow the list by diagnosis code, but it contained all patients for any problem for any physician working in the clinic. The foot ulcer patients comprised approximately 10% of the total list. There are multiple users of patient files and multiple locations required for each file. This made the acquisition of the files somewhat difficult. These difficulties clearly point out a need for an electronic database of patient information.

Other problems encountered and subsequently overcome within the patient files were gaps and inconsistencies. Sometimes missing information existed elsewhere and was found after diligent searching. In other cases, the information did not exist anywhere else. For example, there were times when one visit would conclude that the patient was doing fine; the next visit would mention that the amputation was healing with no intervening data concerning the need for or the scheduling of an amputation. There were surgical reports inserted in the patient records but they contained no information as to why an amputation was performed. Another hurdle had to do with language coding. It is helpful to the physician to be able to add comments, but these comments have to be coded in very standard language for the database. This is a standard problem in the area of medical informatics.

Many of these problems were expected, since patient records cover not only a variety of patient illnesses but are also available to many different physicians attending to sometimes very different aspects of a patient's condition. Another factor is the nature of the patients the clinic is concerned with; diabetic patients have many possible complications due to their underlying illness.

1b. Solutions

Personnel within the clinic were extremely helpful in overcoming the patient record acquisition problems, but three weeks were still needed to get into a routine process which was acceptable to everyone.

The data gaps encountered from time to time were generally one of two types: incomplete or omitted. The incomplete data could usually be found within post-operative notes or subsequent records; there was no recovery possible for the omitted data.

Translation of records into a form suitable for the database was a task that could only be tackled by developing a working knowledge of the necessary medical terms and the physicians' shorthand for that terminology. Again, the clinic personnel proved to be invaluable in regards to this task. Also aiding in this pursuit were the forms that were used on newer patients; these forms narrowed the doctors' responses, and were very helpful with the transition from chart notes to database entries.

The language coding problem was not as enormous a problem as expected, due in large part to the preprinted forms that some of the newer patient files contained; they indicated what was considered essential information on patient condition, and helped structure the language. The development of paper forms is an essential initial step in the development of electronic patient files.

2. Data Summary

2a. General

The following table summarizes the **return** visits for each of the 80 patients:

Summary of the Return Visits				
Type of Visit	Average Number Per Patient	Maximum Number of Visits	Standard Error for the Average	Total Number of Patient Visits
One Week Interval	7.15	62	1.16	572
Two Weeks	4.21	39	0.62	337
One Month	3.09	26	0.43	247
Two Months	4.58	31	0.66	366
More Than Two Months	3.08	17	0.39	246

This is in addition to the 80 initial visits made by the 80 patients sampled. Note that the majority of visits occur at one- or two- week intervals. These patients all have infected ulcers and have the most severe problems. The following table summarizes the amputations by category for the 80 patients:

TYPE	NUMBER
Full toe Amputations	11
Hammer Toe Corrections	5
At Interphalangeal Joint	3
Distal Phalanx Removals	2
Full Below-the-Knee Amputations	2
Bony Resections	2
TOTAL	25

If you consider the fact that there were a total of 1,760 visits (including 44 calls) from these 80 patients, the grand total of 25 amputations is quite remarkable.

Also worth mentioning is the fact that there were 26 surgical debridements, 26 metatarsal head resections, and 7 skin grafts (three were full-thickness). In addition to all this, there were several bypasses and revisions.

2b. Compliance

The question of patient compliance is one that all doctors are concerned with. In the past, patient condition and surveys have been the only indicators of compliance or non-compliance. When addressing this question from within the structure of the database, though, we may use other compliance indicators, such as the difference between when the patient was asked to return to the doctor's office and their actual return date. Consider the following chart, which summarizes the differences between patients' expected return dates and actual return dates:

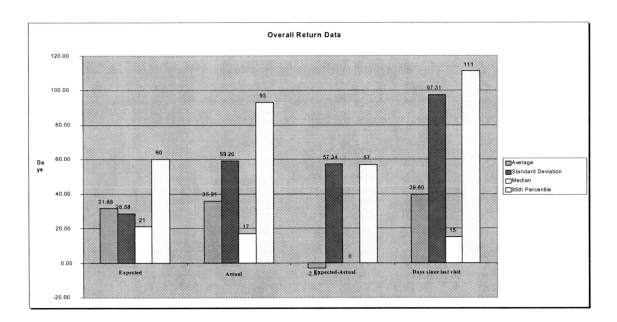

As can be seen from the chart, patients were asked to return in an average of 32 days, and taken as a whole, patients complied with this. But if we were to consider the same data with the results by category of ulcer status, the results are much more useful, as seen on the following charts:

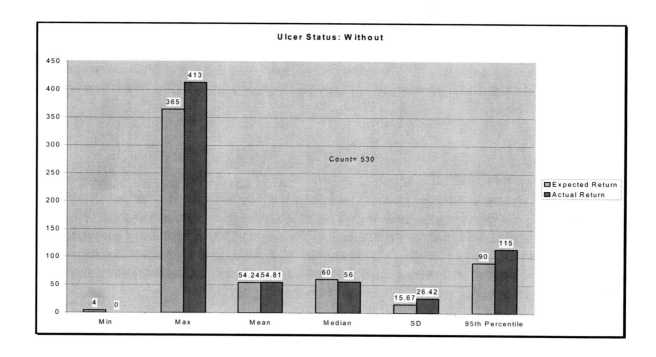

Most of the patients without ulcers complied with anticipated visits to the clinic.

For patients with infected ulcers, there was a median difference of one day between expected and actual return; a difference that does not exist for patients with noninfected ulcers, or without ulcers.

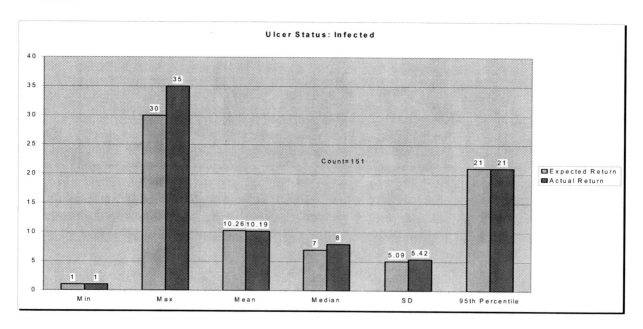

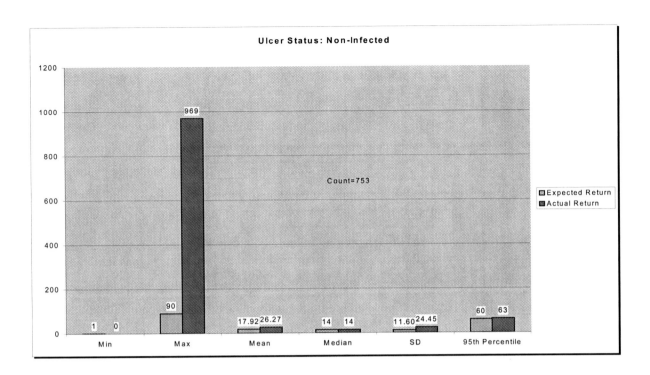

We can now see that the appointment-keeping compliance among those patients with infected ulcers is actually less than that of patients without any ulcers.

This is a somewhat unexpected result; it would be natural to assume that a patient with an infected ulcer would have the highest degree of appointment-keeping compliance. What we have seen, though, indicates that the patients with the highest rates of compliance are those without ulcers at all. This is a good indication that the preventative care at the clinic is not only benefiting the patients, but that they are keenly aware of it as well.

The following table shows the average number and percentage of visits by category of ulcer status:

Ulcer Status:	Unknown	Infected	Non-Infected	Without
Average	5	4	12	9
Avg. (%)	17	8	45	30

Table 1. Average number and percentage of visits by category of ulcer status.

As the table shows, the clinic has seen the 80 patients sampled an average of 30 times, with only 8% of those visits being times when the patient had an infected ulcer. This result is even more remarkable considering that the visits <u>include</u> the initial visit, when the patient almost undoubtedly has infected ulcers.

Also worth noting is the time span between visits. If a patient has an infected ulcer, the time between visits is usually only a few days, but if a patient is without ulcers, the time between visits is usually considerably longer. Of importance also is the distribution of the number of visits. Consider the number of visits per patient at one-week intervals:

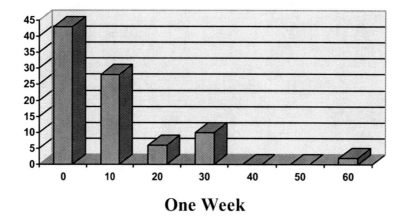

One Week

Note that the graph is exponential. Most of the patients, once started with the clinic, continue. Therefore, those patients with 10 or more visits are long-term. Very few patients discontinue coming to the clinic, although the degree of compliance varies. Many patients who do not have an ulcer tend to increase the length of time between visits. Note also that there is a definite change in behavior once the patients start coming to the clinic. The first visit usually occurs when the patient has an infected ulcer. If the patient has a new ulcer (once the initial ulcer has healed), the next visit occurs before the new ulcer becomes infected.

Another way of measuring compliance would be to give each patient a "grade" of A, B, C, D, or F according to the difference between when they were instructed to return and when they actually did. Consider the following grade scale and the results obtained from using this scale:

Number of days past expected return	Grade
0-4	A
5-9	B
10-19	C
20-29	D
30+	F

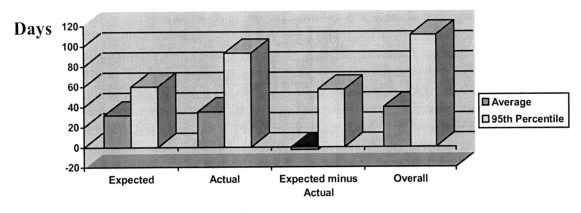

It is easy to see that the average for the actual return is well within 4 days of expected return. This means that as an average, the patients are returning even <u>before</u> the expected return date. This is yet another indicator of the patients' understanding of the benefits of preventative care.

3. Conclusion

The information gathered thus far is useful to both the doctors administering care to these patients and to the patients themselves, although far from complete. Further investigations are needed, and will be done with the aid of more sophisticated statistical techniques and the use of the data mining software.

Abstract for Example 2

Objective: To examine compliance with a hospital protocol for patients with diabetes undergoing open heart surgery.

Method: A total of 223 patient charts were examined retrospectively for compliance with an insulin protocol initiated in 2000. A ranking of compliance levels 1-3 was defined and each patient chart with assigned a ranking.

Results: Full compliance occurred for only 52 out of the 233 patients (23%) with partial compliance in 106 (47%). For 30% of the patients, there was no compliance at all. Infection rates were 3% for the full compliance group compared to 26% for the non compliance group. Glucose levels were less likely to exceed 200 with full compliance ($p<0.0001$).

Conclusion: It is possible to get uncertain results, if the study does not include an examination of compliance with the protocol. With compliance taken into consideration, it can be shown that an insulin protocol can reduce infection and improve outcomes in patients with diabetes.

Introduction

Most clinical studies assume that the defined protocol is performed as defined. This is particularly true of retrospective studies. However, as is often the case, medical staff makes changes to, or does not follow the defined protocol. This lack of adherence to the protocol can bias the results of a statistical study by showing inconclusive results or non-significant results. The purpose of this paper is to demonstrate the difference in outcomes resulting from the inclusion of staff compliance. At issue was a protocol designed to reduce the infection rate of patients with diabetes in open heart surgery. Patients were identified as diabetic prior to surgery and their glucose levels were measured at required intervals, with adjustments to insulin when identified as needed. Initial results did not show a difference in infection rate when patient outcomes were compared to historical data. However, once compliance was ranked on a scale 1-3, infection was significantly reduced in the patients where full compliance was practiced. Kernel density estimation was a tool used to examine the data. Averages also masked outcomes; maximum peak glucose levels were more strongly related to infection rates.

It has been shown that prolonged antibiotic prophylaxis after cardiovascular surgery provides little benefit but some risks.[1] However, benefit would ensue if antibiotic prophylaxis were used in a group of high risk patients. It has been shown that patients with diabetes are more susceptible to mortality and morbidity, including increased risk of infection.[2] Generally, glycemic control is of benefit to patients with diabetes generally, and can help to decrease risk in surgery.[3] Glycemic control can help to reduce the risk of infection, and can help to define the high risk population for antibiotic prophylaxis.[4] Insulin protocols before, during, and after surgery can improve patient outcomes.[5-7] Insulin infusion has been shown to improve white blood cell counts, with the potential of reduction of infection.[8] Protocols have been established for insulin therapy.[9,10] However, there are institutional barriers to the use of an insulin protocol.[11] Because the protocol was established locally, comparison to guidelines was not feasible. Because the protocol was not part of a clinical trial, traditional oversight was not used. However, its relationship to outcome was crucial.

Method

In 2000, a protocol for insulin/glucose treatment was initiated for all patients undergoing CABG surgery with a fasting glucose blood sugar level greater than 126. The primary purpose of the protocol was to reduce infection rates in patients with diabetes, including those who were not previously diagnosed as diabetic. The protocol required blood glucose testing every hour before and after surgery, and every 30 minutes during surgery.

For a 4-month period, 223 patient records were examined retrospectively. The compliance levels with the protocol were ranked as follows:

1 = On protocol / Compliant

2 = Partial complaince: Ins. drip initiated with monitoring & some adjustments made BUT timing was not correct and protocol was not exactly followed &/or not aggressive enough to decrease blood glucose and bolus was not given.

3 = Not Compliant: Did nothing, no glucose records during surgery, skipped more than 2 hrs without glucose readings during OR. Did Insulin bolus only and no drip started

Results

The glucose data contained a ranking of protocol compliance. Using this ranking for only protocol patients (defined with a fasting glucose value of 126 or better), an analysis was performed to determine the relationship between glucose level, compliance level, and infection rate. Table 1 indicates this relationship ($p=0.0008$).

Table 1. Number of patients at full, partial, and no compliance to protocol by infection rates.

Compliance Level	Non-Infection	Infection
Full Compliance	49 (94%)	3 (6%)
Partial Compliance	97 (92%)	9 (8%)
No Compliance	48 (74%)	17 (26%)

Note that there is a statistically significant increase in the infection rate as the compliance level decreases. Figure 1 demonstrates that the maximum glucose level remains within a narrow range from 180 to 280 for full compliance. However, for partial and no compliance, the maximum glucose level can range from 100 to over 300.

Figure 1. Comparison of Protocol Compliance Levels by Patient's Peak Glucose Levels

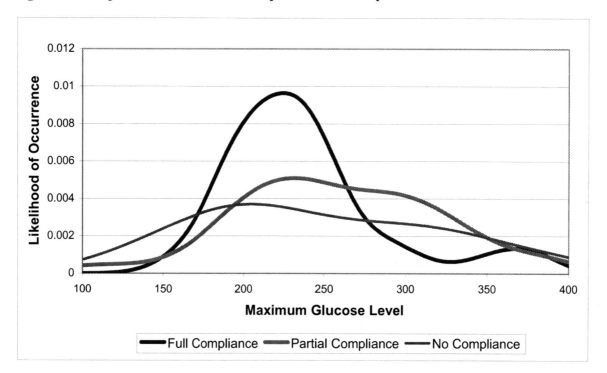

Patients with no compliance were 13 times more likely to have an infection than the full compliance group; patients in the partial compliance group were 9 times as likely to have an infection. An examination of the partial compliance group demonstrates that infected patients are more likely to have higher maximum glucose levels. As noted in Figure 2, patients with infection are more likely to have a maximum value in the range between 220 and 279; patients without infection are more likely to have a peak level at 200.

Figure 2. Comparison of Protocol by Infection for No Compliance Group

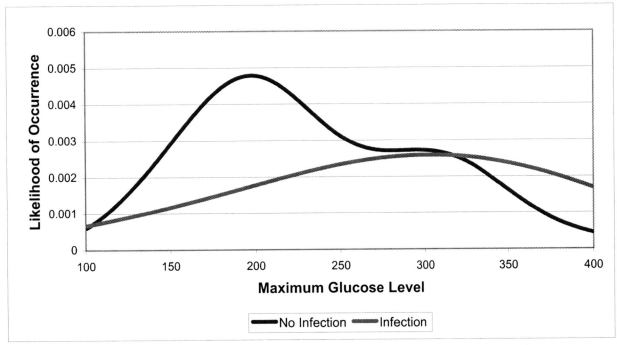

Note that virtually no patient with infection had a peak glucose level less than 200. With partial compliance, the differential in glucose level is also significant ($p<0.0001$, Figure 3).

Figure 3. Comparison of Protocol by Infection for Partial Compliance

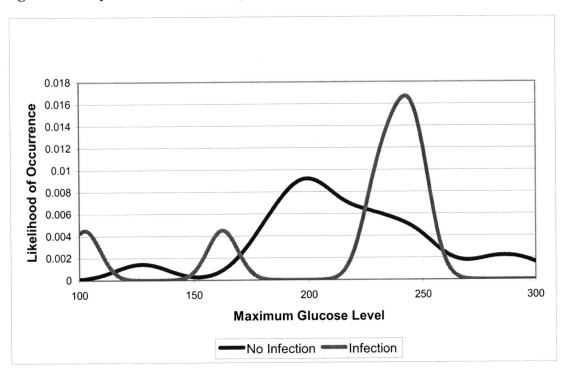

In addition, data for a 4-month period when the protocol called for a 3-day insulin drip after surgery were compared to the data from historical data in 1999 to determine whether the 3-day insulin drip resulted in a further reduction in infections compared to a 1-day insulin drip. Only patients with no compliance were compared (Figure 4) as patients with full compliance had virtually no infection rate (Table 2).

Figure 4. Comparison of 1-day and 3-day Protocols for Full Compliance

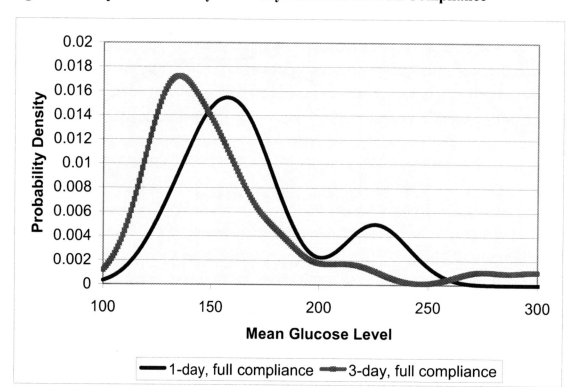

Table 2. Number of patients with 3-day insulin after surgery compared to 1-day protocol.

Protocol	Full Compliance	Partial Compliance	No Compliance
1-day	34 (16%)	15 (7%)	16 (8%)
3-day	33 (20%)	73 (44%)	36 (21%)

The more recent data have a lower overall level of compliance with the protocol. The result is statistically significant (p<0.0001, chi-square). Because of the small numbers of infection, Fisher's Exact Test was used to determine statistical differences between the two sets of data. Overall, the 1-day protocol had an infection rate of 9%; the 3-day protocol had a rate of 2% (p=0.0131). Restricting the data to the patients with full compliance to the protocol, the rates were 3% (1-day) and 6% (3-day) with a p-value of 0.6135. It should be noted that the small number of patients in this category yields insufficient power to declare that there is no difference in the rate. Similarly, there is insufficient power to compare the partial compliant patients (7%,

1-day; 0%, 3-day; p-value=0.1705). For patients with no compliance, the overall infection rate is high (25% 1-day; 14%, 3-day; p=0.4312).

With the 1-day protocol, there is a higher probability that the maximum glucose level will exceed 270 when compared to the 3-day protocol. With no compliance, the difference is a slight shift in mean glucose levels. Note that there is a statistically significant increase in the infection rate as the compliance level decreases.

Figure 5. Comparison of 1-day and 3-day Protocols for Partial Compliance

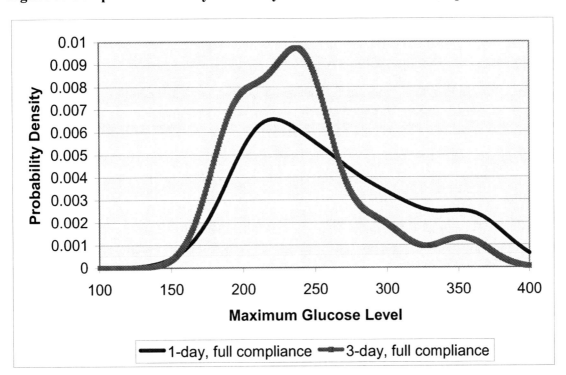

Conclusion

The effectiveness of a protocol in improving patient outcomes must be examined in relationship to compliance to the protocol. If the hospital staff has poor compliance, the outcomes cannot be related to the protocol itself. However, if it is generally assumed that a protocol is routinely followed without examination to validate that fact, outcomes may become inconclusive, and the inconclusive result can create pressure to discontinue the protocol even though it might actually be effective when followed.

References

1. Harbarth S, Samore M, Lictenberg D, Carmeli Y. Prolonged antibiotic prophylaxis after cardiovascular surgery and its effect on surgical site infections and antimicrobial resistance. *Circulation.* 2000;101(25):2916-2921.

2. Cohen Y, Raz I, Merin G, Mozes B. Comparison of factors associated with 30-day mortality after coronary artery bypass grafting in patients with versus without diabetes mellitus. *The American Journal of Cardiology.* 1998;81(1):7-11.

3. Webster MWI, Scott RS. What cardiologists need to know about diabetes. *The Lancet.* 1997;250(9072S):23sI-28sI.

4. Trick WE, Scheckler WE, Tokars J, et al. Modifiable risk factors associated with deep sternal site infection after coronary artery bypass grafting. *The Journal of Thoracic & Cardiovascular Surgery.* 2000;119(1):108-114.

5. Jacober SJ, Sowers JR. An update on perioperative management of diabetes. *Archives of Internal Medicine.* 1999;159(20):2405-2411.

6. Kalin MF, Tranbaugh RF, Salas J, et al. Intensive intervention by a diabetes team diminishes excess hospital mortality in patients with diabetes who undergo coronary artery bypass graft. *Diabetes.* 1998;47(1S):87A.

7. Lazar HL, Philippides G, Fitzgerald C, Lancaster D, Shemin RJ, Apstein C. Glucose-insulin-potassium solutions enhance recovery after urgent coronary artery bypass grafting. *The Journal of Thoracic & Cardiovascular Surgery.* 1997;113(2):354-362.

8. Rassias AJ, Marrin CAS, Arruda J, Whalen PK, Beach M, Yeager M. Insulin infusion improves neutrophil function in diabetic cardiac surgery patients. *Anesthesia & Analgesia.* 1999;88(5):1011-1016.

9. Markovitz L, Wiechmann R, Harris N, et al. Description and evaluation of a glycemic management protocol for patients with diabetes undergoing heart surgery. *Endocrine Practice.* 2002;8(1):10-18.

10. Carlson G. The value of point-of-care testing when instituting an insulin drip protocol. *Critical Care Nursing Quarterly.* 2001;24(1):49-53.

11. Metchick L, Jr WAP, Inzucchi S. Inpatient management of diabetes mellitus. *The American Journal of Medicine.* 2002;113(4):317-323.

4.7. Exercises

1. Given the hypotheses you formulated in Chapter 3 concerning the student survey, use table analysis to investigate the student responses.

Chapter 5. Basics of Inference for Categorical Data

5.1. Introduction 159

5.2. Data Summary 160

5.3. Discussion of the Problem 165

5.4. Chi Square Analysis 180

5.5. Logistic Regression 199

5.6. Exercises 212

5.1. Introduction

The methods we focus on in the chapter are chi-square and logistic regression. A chi square test compares two categorical variables. The hypothesis tested is

H_0: there is no relationship between the two variables

H_1: Some relationship exists between the two variables

If H_0 is rejected, there is some relationship between the two variables. If we can't reject H_0, we can't necessarily accept H_0 and declare that no relationship exists. We just can't prove that one exists.

Logistic regression compares one outcome variable Y to several input variables written as a linear equation. That is,

$$Y = \beta_0 + \beta_1 X_1 + \beta_2 X_2 + ... + \beta X_n$$

where Y is a categorical variable. The X variables can be of any type. The first hypothesis tested is:

H_0: The variability in Y has no relationship to the variability in the X variables

H_1: Some relationship exists between Y and the X variables

If we reject H_0, then we can examine the X-variables in more detail to determine which of the X-variables are related to Y. That is, we test:

H_0: X_i has no relationship to Y

H_1: X_i has some relationship to Y

Depending upon the distribution of Y, we can modify the expected shape of the linear function,

$$\beta_1 X_1 + \beta_2 X_2 + ... + \beta X .$$

We will discuss this model in more detail later in the chapter.

The Medical Expenditure Panel Survey contains information for a cohort of individuals concerning their use of medical services. These datafiles are posted at http://www.ahrq.gov/data/mepsix.htm. Data can be downloaded from this site for use in research. In this chapter, we discuss the inferential methodology in the context of examining these data. We use dataset numbered HC85a. It contains information about medication prescriptions for the year 2001.

5.2. Data Summary

Before performing inference, it is helpful to examine the data using the visualization and summary techniques to get a good idea of the information contained in the data. One way to do this quickly is to use the Characterize option in Enterprise Guide (Display 5.1).

Display 5.1. Characterize option

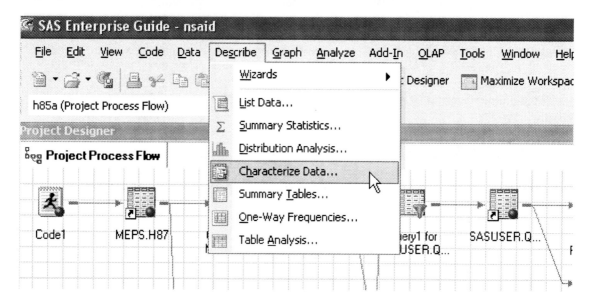

It gives three screens. The first screen identifies the dataset or datasets to be summarized. The second screen allows you to specify which summaries to perform (Display 5.2).

Display 5.2. Screen to Specify Datasets

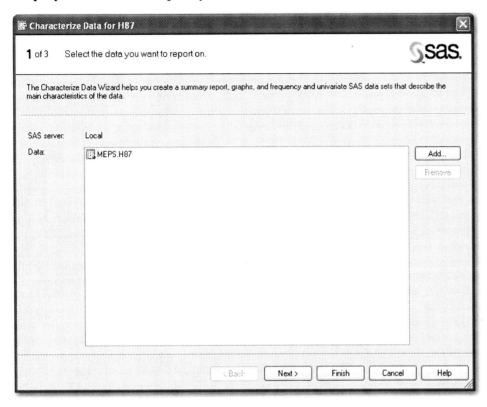

Display 5.3. Screen to Select Report Options

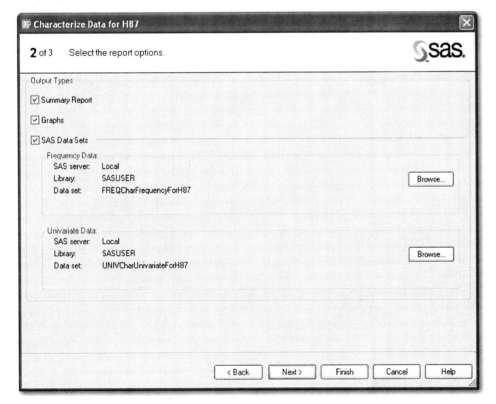

Display 5.4. Screen to Limit Categorical Levels

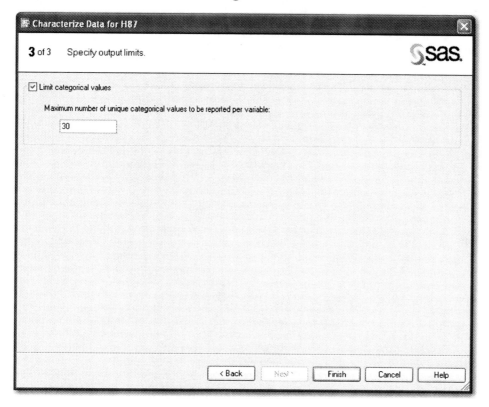

To ensure that the summaries are performed on a timely basis, you can restrict the number of categorical levels reported. The default is to limit reporting to 30 levels.

The following results are returned in the Project Diagram (Display 5.5).

Display 5.5. Project Diagram for Characterize Option

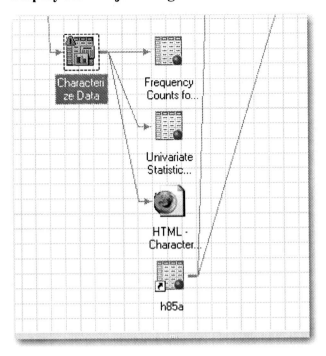

Note that frequencies and summaries are returned as datasets (Display 5.6).

Display 5.6. Frequency Count Dataset

	DataSet	Variable	Label	Format	Value	Count	Percent
1	MEPS.H87	CCCODEX	CLINICAL CLASS		126	8700	8.1507991531
2	MEPS.H87	CCCODEX	CLINICAL CLASS		098	4889	4.5803743746
3	MEPS.H87	CCCODEX	CLINICAL CLASS		134	3852	3.6088365905
4	MEPS.H87	CCCODEX	CLINICAL CLASS		204	3585	3.3586913751
5	MEPS.H87	CCCODEX	CLINICAL CLASS		205	3346	3.1347786168
6	MEPS.H87	CCCODEX	CLINICAL CLASS		135	3154	2.9548989114
7	MEPS.H87	CCCODEX	CLINICAL CLASS		053	2760	2.5857707658
8	MEPS.H87	CCCODEX	CLINICAL CLASS		658	2746	2.5726545373
9	MEPS.H87	CCCODEX	CLINICAL CLASS		259	2504	2.3459311585
10	MEPS.H87	CCCODEX	CLINICAL CLASS		211	2286	2.141692743
11	MEPS.H87	CCCODEX	CLINICAL CLASS		651	2185	2.0470685229
12	MEPS.H87	CCCODEX	CLINICAL CLASS		049	2079	1.9477599355
13	MEPS.H87	CCCODEX	CLINICAL CLASS		200	2059	1.9290224662
14	MEPS.H87	CCCODEX	CLINICAL CLASS		-1	1949	1.825966385
15	MEPS.H87	CCCODEX	CLINICAL CLASS		128	1835	1.7191628099
16	MEPS.H87	CCCODEX	CLINICAL CLASS		092	1709	1.6011167532
17	MEPS.H87	CCCODEX	CLINICAL CLASS		084	1705	1.5973692593
18	MEPS.H87	CCCODEX	CLINICAL CLASS		255	1608	1.5064925331
19	MEPS.H87	CCCODEX	CLINICAL CLASS		232	1594	1.4933763046

Display 5.7. Data Summary Dataset

	DataSet	Variable	Label	Format	N	NMiss	Total	Min	Mean	Median
1	MEPS.H87	ACCDENTD	DATE OF ACCID		106738	0	-62619	-9	-0.586660796	
2	MEPS.H87	ACCDENTM	DATE OF ACCID		106738	0	-74141	-9	-0.694607356	
3	MEPS.H87	ACCDENTY	DATE OF ACCID		106738	0	13438476	-9	125.90151586	
4	MEPS.H87	ACCDNWRK	DID ACCIDENT O		106738	0	-91251	-9	-0.854906406	
5	MEPS.H87	ACDNTLOC	WHERE DID ACC		106738	0	-29584	-9	-0.277164646	
6	MEPS.H87	ACDNTOTH	WAS SOMETHIN		106738	0	-89205	-9	-0.835737975	
7	MEPS.H87	CONDBEGD	DATE CONDITIO		106738	0	-152167	-9	-1.425612247	
8	MEPS.H87	CONDBEGM	DATE CONDITIO		106738	0	-137849	-9	-1.291470704	
9	MEPS.H87	CONDBEGY	DATE CONDITIO		106738	0	33451054	-9	313.39404898	
10	MEPS.H87	CONDN	CONDITION NUM		106738	0	5558770	10	52.078641159	
11	MEPS.H87	CONDRN	CONDITION ROU		106738	0	246707	1	2.3113324214	
12	MEPS.H87	CRND1	HAS CONDITION		106738	0	-23099	-1	-0.216408402	
13	MEPS.H87	CRND2	HAS CONDITION		106738	0	-20598	-1	-0.192977196	
14	MEPS.H87	CRND3	HAS CONDITION		106738	0	54212	0	0.5078978433	
15	MEPS.H87	CRND4	HAS CONDITION		106738	0	-26929	-1	-0.252290656	
16	MEPS.H87	CRND5	HAS CONDITION		106738	0	-30154	-1	-0.282504825	
17	MEPS.H87	DROWN	WAS DROWNING		106738	0	-86694	-9	-0.812213083	
18	MEPS.H87	ERNUM	# ER EVENTS AS		106738	0	7589	0	0.0710993273	
19	MEPS.H87	FALL	WAS IT A FALL		106738	0	-88707	-9	-0.831072345	
20	MEPS.H87	FIREBURN	WAS FIRE/BURNI		106738	0	-86763	-9	-0.812859525	
21	MEPS.H87	FOLOCA1	RD1: RCV FOLLO		106738	0	-86168	-9	-0.807285128	
22	MEPS.H87	FOLOCA2	RD2: RCV FOLLO		106738	0	-96216	-1	-0.901422174	
23	MEPS.H87	FOLOCA3	RD3: RCV FOLLO		106738	0	-93679	-9	-0.877653694	
24	MEPS.H87	FOLOCA4	RD4: RCV FOLLO		106738	0	-99635	-1	-0.933453878	
25	MEPS.H87	FOLOCA5	RD5: RCV FOLLO		106738	0	-102343	-1	-0.958824411	

Graphics are also provided. The one shown here is for patient condition (Figure 5.1).

Figure 1. Graph of Patient Condition

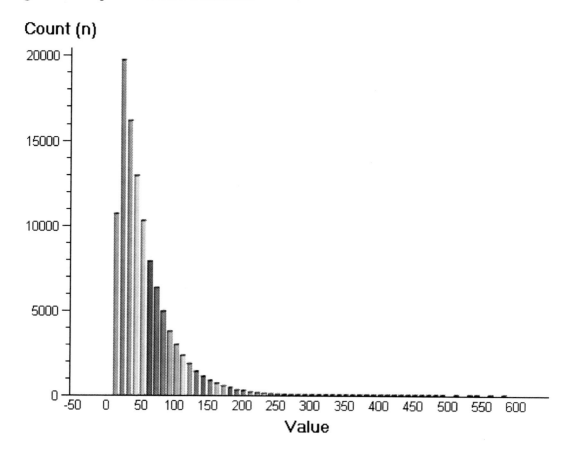

5.3. Discussion of the Problem

We want to consider just part of the data rather than all of it. Patients who have congestive heart failure are told to avoid non-steroidal anti-inflammatory drugs such as aspirin, ibuprofen, Aleve, and so on. The only exception is the use of low dose aspirin. Patients with arthritis, especially rheumatoid arthritis are usually prescribed non-steroidal anti-inflammatory drugs (NSAIDs). What about patients who have both problems? What medications are they prescribed? What are the costs of the medications?

In order to solve this problem, we need to use the filter and querying options in SAS Enterprise Miner to reduce the dataset to the drugs in question. That means using the services of a pharmacist to determine which of the drugs are NSAIDs. The filter and query option is located in the Data menu.

Display 5.8. Filter and Query in Menu

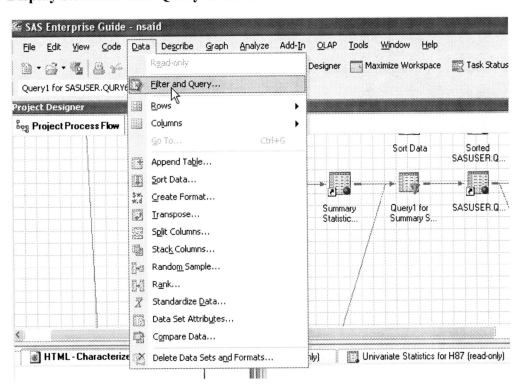

To use the query, we first drag the variables we want to use from the left side to the right side window (Display 5.9).

Display 5.9. Screen to Define Query

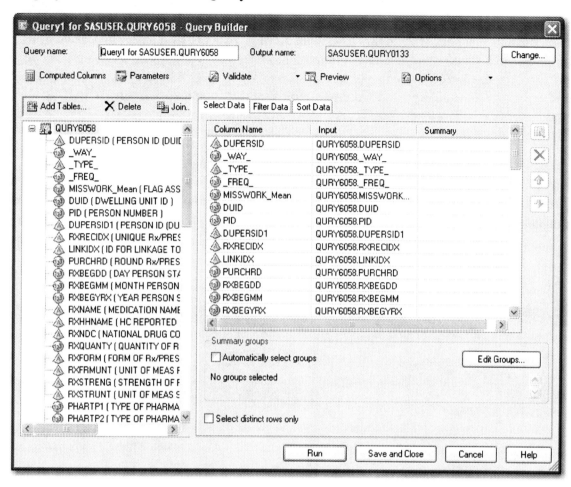

Next, we use the Filter Data tab to restrict the medication names to the NSAIDs. In the Filter Data tab, we drag the variable, "RXNAME" from the left to the right window. Then we get a new window (Display 5.10).

Display 5.10. Filter RXNAME

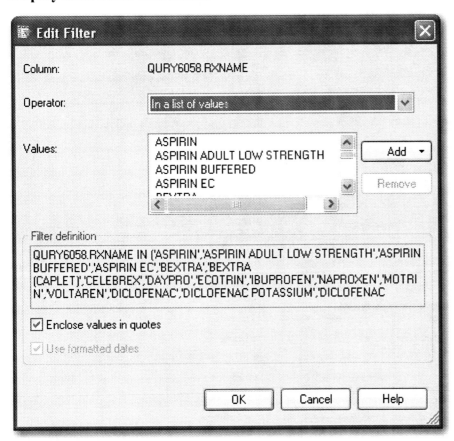

We use the Add button to get a list of values and then use the "Ctrl" button with the mouse to highlight only the drugs that are NSAIDs. Once this query is run, the dataset is reduced to 413 records from the total of over 300,000 records initially.

The next step is to determine just why this medication was prescribed. To find out, we do a frequency count of the diagnosis variable, RXICD1X (Display 5.11).

Display 5.11. Menu for Frequency Counts

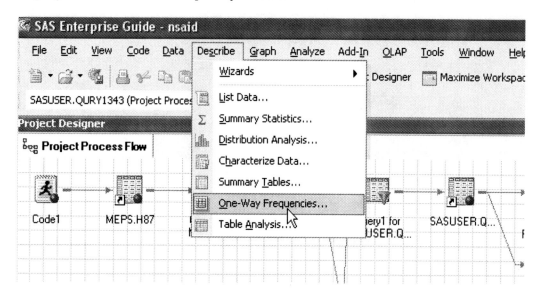

Table 5.1. Results of Frequency Procedure

RXICD1X	Frequency	Percent	Cumulative Frequency	Cumulative Percent
-1	30	7.26	30	7.26
250	6	1.45	36	8.72
272	1	0.24	37	8.96
274	11	2.66	48	11.62
366	1	0.24	49	11.86
401	20	4.84	69	16.71
410	2	0.48	71	17.19
414	2	0.48	73	17.68
428	88	21.31	161	38.98
429	35	8.47	196	47.46
530	6	1.45	202	48.91
701	2	0.48	204	49.39
715	22	5.33	226	54.72
716	162	39.23	388	93.95

RXICD1X	Frequency	Percent	Cumulative Frequency	Cumulative Percent
719	1	0.24	389	94.19
721	3	0.73	392	94.92
727	1	0.24	393	95.16
729	10	2.42	403	97.58
737	1	0.24	404	97.82
753	2	0.48	406	98.31
780	1	0.24	407	98.55
807	2	0.48	409	99.03
840	1	0.24	410	99.27
V68	3	0.73	413	100.00

The table gives the diagnosis codes that correspond to the prescriptions. The most are for the codes 428 and 716. We need to know what these codes represent. Translations for these codes are available online at http://icd9cm.chrisendres.com/index.php?action=contents. The code, 716, represents arthritis of an unspecified type and code 428 represents heart failure. Similarly, 429=complications of heart disease and 715= Osteoarthrosis. Rheumatoid arthritis=714, for which no prescriptions for NSAIDS are included.

The next step in looking at the dataset is to find all patients with arthritis and all patients with heart failure, and to determine the proportion with prescriptions for NSAIDS and those without prescriptions for NSAIDS. To do this, we must do some more filtering on the original dataset. We first find all patients with a diagnosis of heart failure or arthritis for any of the three diagnosis codes, duplicating the filter for each code (Display 5.12).

Display 5.12. Filtering Patient Conditions

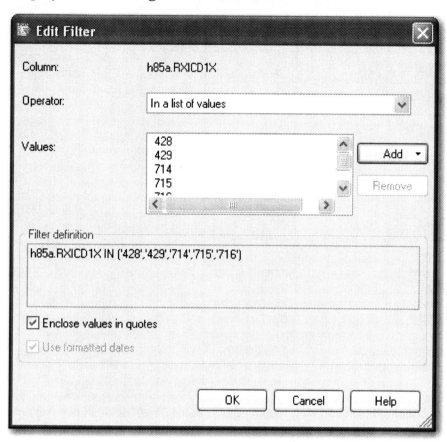

Display 5.13. Filtering Multiple Variables

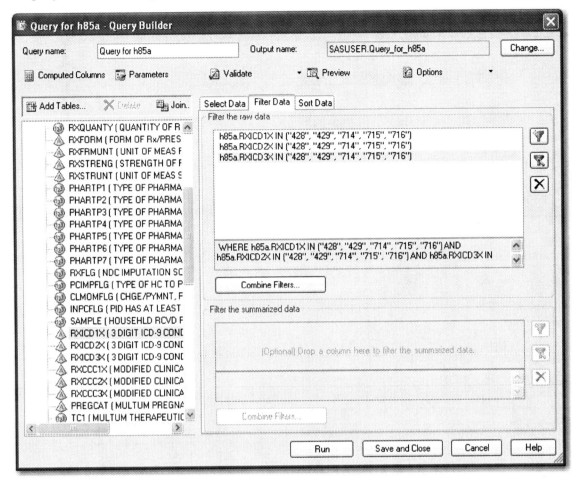

Next, use the Combine Filters button to change the "and" connector to "or" (Display 5.14).

Display 5.14. Changing Connector to "OR"

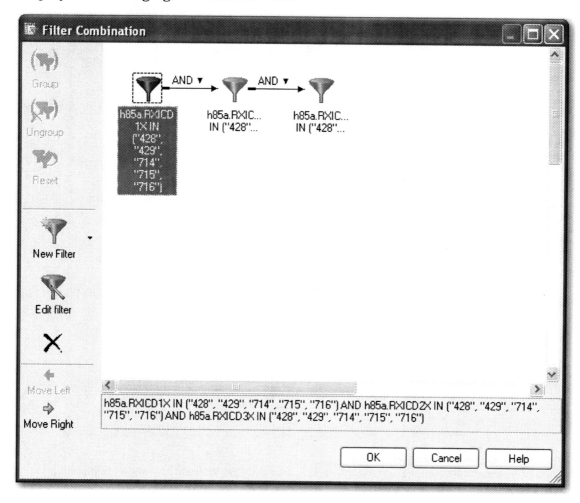

Right-click on the "And" to get "OR" (Display 5.15).

Display 5.15. Change to "OR"

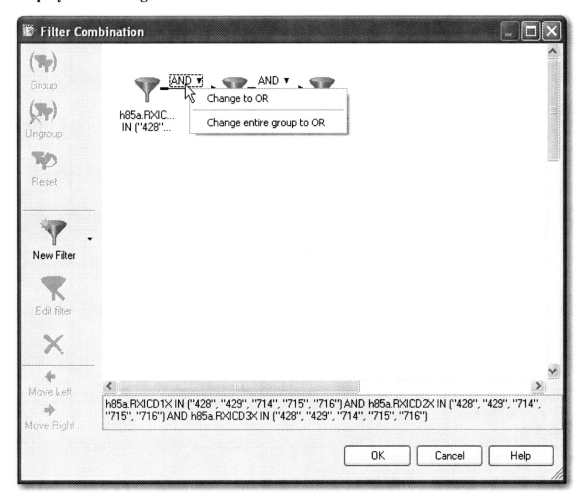

We want to change the entire group to "OR".

Display 5.16. Result of Change

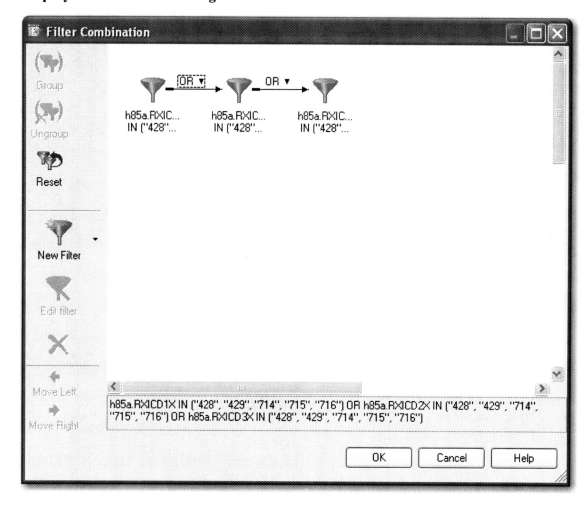

This gives us a total of 19,159 patients. The next step is to separate out who is getting NSAIDS and who is not. The easiest way is to merge the two filtered datasets by patient identifier. We first sort the two datasets, the one containing the heart patients, and the one containing the nsaids.

Display 5.17. Sort Data

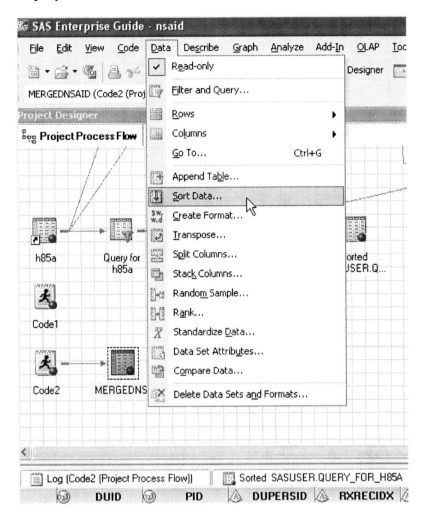

Note the names of these sorted datasets. We sort by the following two variables (Display 5.18).

Display 5.18. Sort by Patient Identifier and by Drug Name

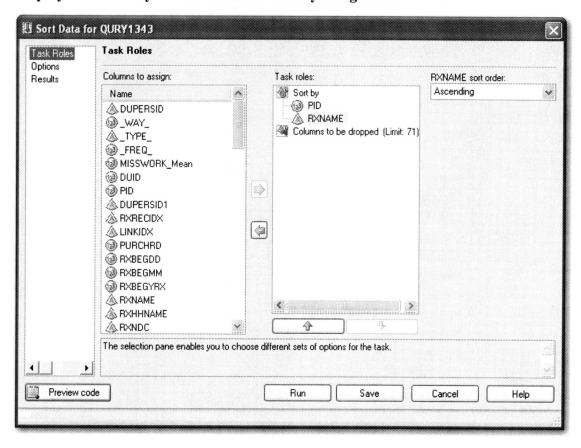

Then we use a code node to perform the following merge of the datasets:

```
data sasuser.mergednsaid;
set sasuser.sortsortedquery_for_h85a
sasuser.sortsortedqury1343;
by PID rxname;
run;
```

for a total of 19,572 observations in the dataset. The last column will be blank when the medication is not for an nsaid. We need to make one more change to the dataset. We want to use the computed column button to recode a column (Display 5.19).

Display 5.19. Compute a Column

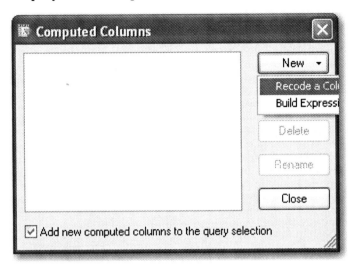

This gives use the following screen (Display 5.20).

Display 5.20. Choice to Recode Column

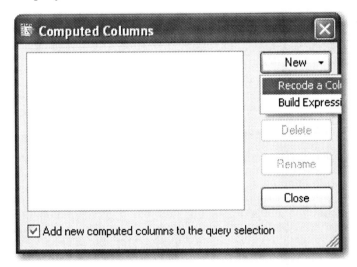

First, we replace all non-blank values with the code of '1' (Display 5.21).

Display 5.21. Choose Values

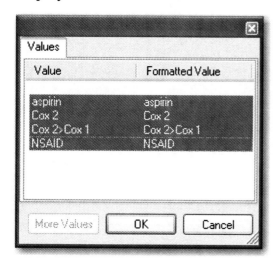

Display 5.22. Change Coding

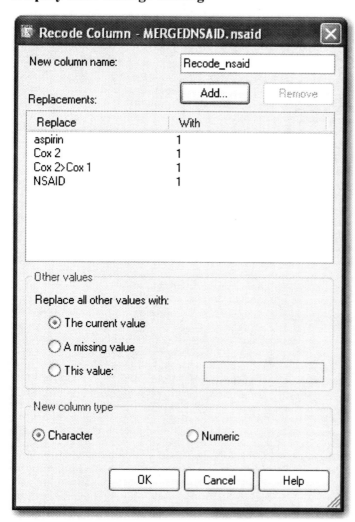

Next, we replace the blanks with the value, "0". As a last step, we rename the variable by clicking on the "Rename" Button.

Display 5.23. Rename Variable

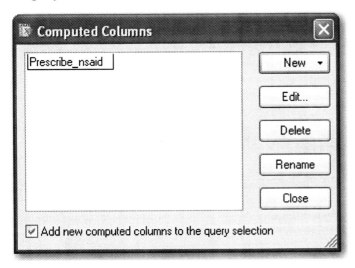

5.4. Chi Square Analysis

To perform a chi-square test between the first diagnosis and the prescription for NSAIDs, we use the table analysis. The hypothesis tested is

H_0: There is no relationship between diagnosis and NSAID Prescriptions

H_1: Some relationship exists

Display 5.24. Table Analysis Screen

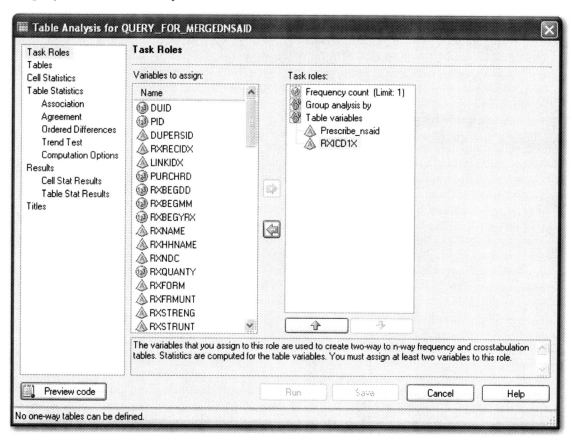

In the tables window, we drag and drop the variables from the right to the left. The first one dragged defines the columns; the next one defines the rows.

Display 5.25. Build Table

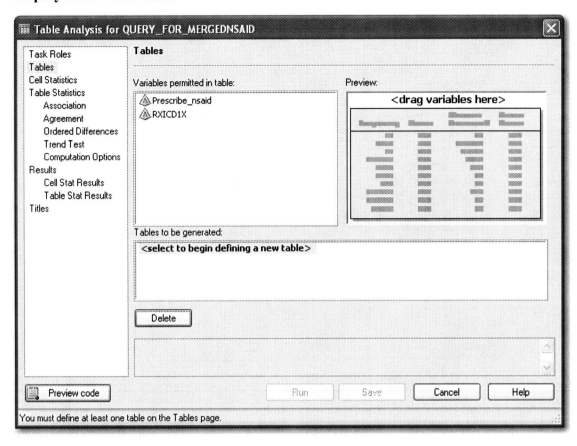

For our problem, we want it to look that in display 5.26. First, drag and drop the row variable; then drag and drop the column variable.

Display 5.26. Defining Table

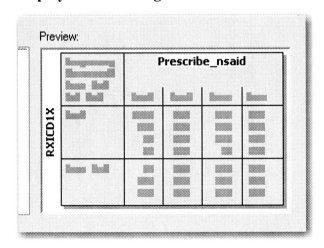

We next go to the cell statistic box to check the "row percentages"

Display 5.27. Including Row Percentages

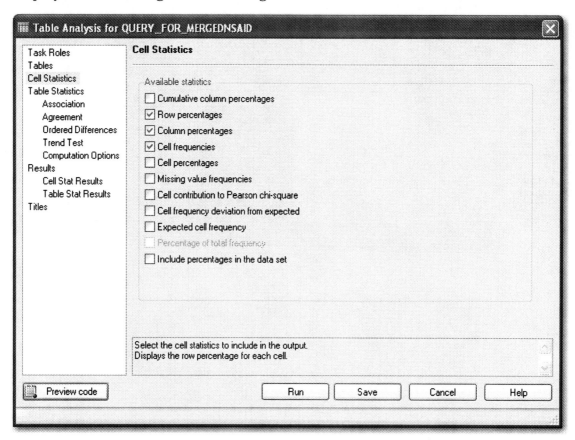

We go to the association box to check the Chi square box (Display 5.28).

Display 5.28. Including Chi Square Statistic

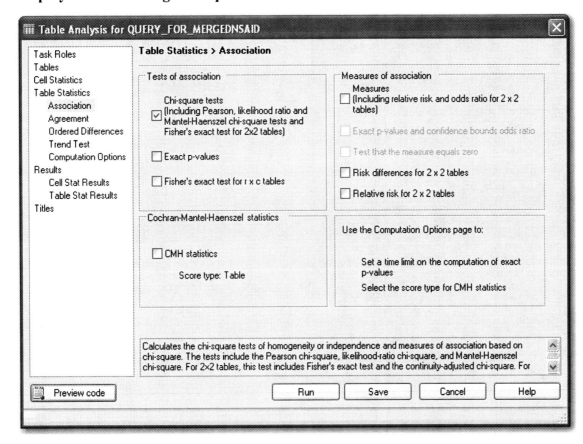

The table immediately shows a problem in that we only want people with heart failure or arthritis or both (Table 5.2).

Table 5.2. Results of Table Analysis

RXICD1X	Prescribe_nsaid		Total
	0	1	
-1	0 0.00 0.00	30 100.00 7.26	30
070	4 100.00 0.02	0 0.00 0.00	4

078	5 100.00 0.03	0 0.00 0.00	5
135	3 100.00 0.02	0 0.00 0.00	3
153	8 100.00 0.04	0 0.00 0.00	8
174	6 100.00 0.03	0 0.00 0.00	6
250	36 85.71 0.19	6 14.29 1.45	42
272	6 85.71 0.03	1 14.29 0.24	7
274	40 78.43 0.21	11 21.57 2.66	51
276	6 100.00 0.03	0 0.00 0.00	6
296	9 100.00 0.05	0 0.00 0.00	9
300	5 100.00 0.03	0 0.00 0.00	5
311	6 100.00 0.03	0 0.00 0.00	6

331	12 100.00 0.06	0 0.00 0.00	12
333	3 100.00 0.02	0 0.00 0.00	3
340	8 100.00 0.04	0 0.00 0.00	8
344	2 100.00 0.01	0 0.00 0.00	2
346	11 100.00 0.06	0 0.00 0.00	11
349	12 100.00 0.06	0 0.00 0.00	12
354	8 100.00 0.04	0 0.00 0.00	8
355	25 100.00 0.13	0 0.00 0.00	25
362	17 100.00 0.09	0 0.00 0.00	17
366	0 0.00 0.00	1 100.00 0.24	1
401	392 95.15 2.05	20 4.85 4.84	412

410	19 90.48 0.10	2 9.52 0.48	21
411	1 100.00 0.01	0 0.00 0.00	1
413	21 100.00 0.11	0 0.00 0.00	21
414	2 50.00 0.01	2 50.00 0.48	4
424	4 100.00 0.02	0 0.00 0.00	4
426	2 100.00 0.01	0 0.00 0.00	2
427	55 100.00 0.29	0 0.00 0.00	55
428	2577 96.70 13.45	88 3.30 21.31	2665
429	5732 99.39 29.92	35 0.61 8.47	5767
433	10 100.00 0.05	0 0.00 0.00	10
444	1 100.00 0.01	0 0.00 0.00	1

459	10 100.00 0.05	0 0.00 0.00	10
477	1 100.00 0.01	0 0.00 0.00	1
486	2 100.00 0.01	0 0.00 0.00	2
492	7 100.00 0.04	0 0.00 0.00	7
493	33 100.00 0.17	0 0.00 0.00	33
496	8 100.00 0.04	0 0.00 0.00	8
505	2 100.00 0.01	0 0.00 0.00	2
511	3 100.00 0.02	0 0.00 0.00	3
518	14 100.00 0.07	0 0.00 0.00	14
530	1 14.29 0.01	6 85.71 1.45	7
593	4 100.00 0.02	0 0.00 0.00	4

623	2 100.00 0.01	0 0.00 0.00	2
701	0 0.00 0.00	2 100.00 0.48	2
710	26 100.00 0.14	0 0.00 0.00	26
714	1290 100.00 6.73	0 0.00 0.00	1290
715	728 97.07 3.80	22 2.93 5.33	750
716	7282 97.82 38.01	162 2.18 39.23	7444
717	4 100.00 0.02	0 0.00 0.00	4
719	54 98.18 0.28	1 1.82 0.24	55
721	7 70.00 0.04	3 30.00 0.73	10
722	40 100.00 0.21	0 0.00 0.00	40
723	33 100.00 0.17	0 0.00 0.00	33

724	107 100.00 0.56	0 0.00 0.00	107
726	8 100.00 0.04	0 0.00 0.00	8
727	23 95.83 0.12	1 4.17 0.24	24
728	6 100.00 0.03	0 0.00 0.00	6
729	153 93.87 0.80	10 6.13 2.42	163
733	43 100.00 0.22	0 0.00 0.00	43
737	19 95.00 0.10	1 5.00 0.24	20
753	0 0.00 0.00	2 100.00 0.48	2
756	17 100.00 0.09	0 0.00 0.00	17
780	26 96.30 0.14	1 3.70 0.24	27
782	10 100.00 0.05	0 0.00 0.00	10

784	17 100.00 0.09	0 0.00 0.00	17
785	9 100.00 0.05	0 0.00 0.00	9
786	8 100.00 0.04	0 0.00 0.00	8
790	13 100.00 0.07	0 0.00 0.00	13
807	0 0.00 0.00	2 100.00 0.48	2
820	4 100.00 0.02	0 0.00 0.00	4
825	1 100.00 0.01	0 0.00 0.00	1
831	3 100.00 0.02	0 0.00 0.00	3
836	1 100.00 0.01	0 0.00 0.00	1
839	3 100.00 0.02	0 0.00 0.00	3
840	12 92.31 0.06	1 7.69 0.24	13

959	4 100.00 0.02	0 0.00 0.00	4
995	2 100.00 0.01	0 0.00 0.00	2
V42	12 100.00 0.06	0 0.00 0.00	12
V43	4 100.00 0.02	0 0.00 0.00	4
V45	14 100.00 0.07	0 0.00 0.00	14
V53	1 100.00 0.01	0 0.00 0.00	1
V58	2 100.00 0.01	0 0.00 0.00	2
V68	5 62.50 0.03	3 37.50 0.73	8
V77	1 100.00 0.01	0 0.00 0.00	1
V81	32 100.00 0.17	0 0.00 0.00	32
Total	19159	413	19572

We can correct this problem by using the compute column button. We want to code a 1 if the condition is 428,429 or a value of -1 if it is one of 714,715,716 and a 0 otherwise for each of the three diagnoses columns (Display 5.29).

Display 5.29. Recode Values to Reduce Table

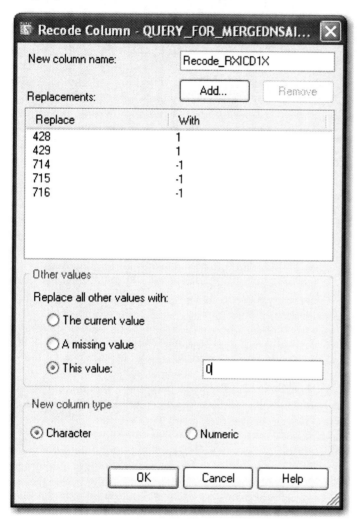

In this case, we define three tables (Display 5.30).

Display 5.30. Defining Multiple Tables

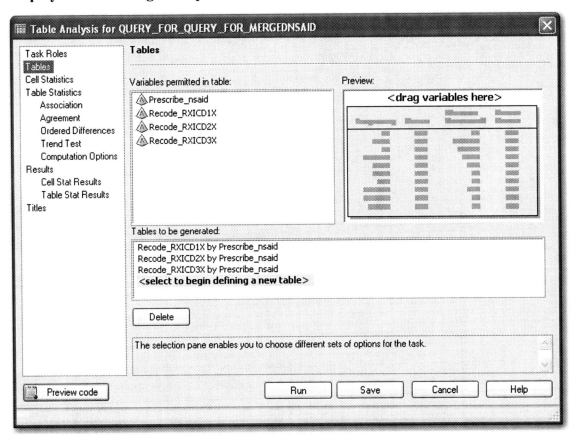

Table 5.3. Results of First Table Analysis from Display 5.30

Frequency Row Pct Col Pct	Table of Recode_RXICD1X by Prescribe_nsaid			
	Recode_RXICD1X	Prescribe_nsaid		Total
		0	1	
	-1	9300 98.06 48.54	184 1.94 44.55	9484
	0	1550 93.60 8.09	106 6.40 25.67	1656
	1	8309 98.54 43.37	123 1.46 29.78	8432
	Total	19159	413	19572

There is a clear difference here in that 98% of those with arthritis are prescribed NSAIDS but only 94% are prescribed for other categories. However, 98% of patients with heart failure are also prescribed NSAIDs. We test the hypothesis

H_0:probability of NSAID for heart failure=probability of NSAID for arthritis=probability of NSAID for all other problems

H_1: at least two probabilities are different (ie, there is a relationship between NSAID and patient condition)

SAS provides three different chi-square statistics; usually they have the same conclusion. In this case, the probability of being wrong when rejecting H_0 is less than 0.0001. Therefore, we can be fairly confident of rejecting the null hypothesis in favor of the alternate. The table indicates that the prescription rate for NSAIDs is lower for all other problems.

Table 5.4. Chi Square Statistic

Statistic	DF	Value	Prob
Chi-Square	2	166.2501	<.0001
Likelihood Ratio Chi-Square	2	117.0662	<.0001
Mantel-Haenszel Chi-Square	1	4.0808	0.0434
Phi Coefficient		0.0922	
Contingency Coefficient		0.0918	
Cramer's V		0.0922	

Table 5.5. Table Analysis for Second Patient Condition

Frequency Row Pct Col Pct	Table of Recode_RXICD2X by Prescribe_nsaid		
Recode_RXICD2X	Prescribe_nsaid		Total
	0	1	
-1	1039 99.33 5.42	7 0.67 1.69	1046
0	17586 97.78 91.79	399 2.22 96.61	17985
1	534 98.71 2.79	7 1.29 1.69	541
Total	19159	413	19572

Again, there is a difference, but the difference is statistically significant. One of the reasons for this is that the larger the sample size, the smaller the difference that is statistically significant. Also, a prescription for NSAID is small compared to all non-NSAID prescriptions.

Table 5.6. Chi Square Test

Chi-Square	2	13.2814	0.0013
Likelihood Ratio Chi-Square	2	17.3143	0.0002
Mantel-Haenszel Chi-Square	1	3.4925	0.0616
Phi Coefficient		0.0260	
Contingency Coefficient		0.0260	
Cramer's V		0.0260	

There is a relationship between patient condition and prescriptions for NSAIDs. If the third diagnosis is for heart failure or arthritis, none of the prescriptions are for NSAIDs (Table 5.7).

Table 5.7. Analysis for Third Patient Condition

Frequency Row Pct Col Pct	Table of Recode_RXICD3X by Prescribe_nsaid			
	Recode_RXICD3X	Prescribe_nsaid		Total
		0	1	
	-1	265 100.00 1.38	0 0.00 0.00	265
	0	18805 97.85 98.15	413 2.15 100.00	19218
	1	89 100.00 0.46	0 0.00 0.00	89
	Total	19159	413	19572

Table 5.8. Chi Square Analysis

Statistic	DF	Value	Prob
Chi-Square	2	7.7715	0.0205
Likelihood Ratio Chi-Square	2	15.2395	0.0005
Mantel-Haenszel Chi-Square	1	1.8946	0.1687
Phi Coefficient		0.0199	
Contingency Coefficient		0.0199	
Cramer's V		0.0199	

Again, there is a relationship between NSAID use and patient condition.

5.5. Logistic Regression

Another way of looking at the data, using all three diagnostic codes is to use logistic regression. The standard procedure for logistic regression has been to use an equation of the form

$$y = B_0 + B_1x_1 + \ldots + B_kx_k + e$$

where y = dependent variable (variable to be modeled – sometimes called the response variable), $x_1, x_2, \ldots, x_k$ = independent variable (variable used as a predictor of y), e = random error, and B_i determines the contribution of the independent variable x_i. The variable y can be continuous (such as length of stay or costs), or discrete (such as mortality). Each x variable denotes the presence or absence of a factor for a particular observation. Therefore, y is equal to the sum of the weights B_i for each factor x_i that is positive for that patient. If y is a continuous variable such as length of stay, then the predicted LOS is the linear combination of weights. If y is a discrete variable such as mortality, an optimum threshold value of found, and mortality is predicted if the sum of the weights exceeds that threshold. In considering the NSAIDs, Y is discrete with two possible values.

Consider an example where there are five patient codes (A,B,C,D,E) used in a logistic regression with the outcome variable of using an NSAID. Then the logistic regression equation can be written

$$P = \alpha_0 + \alpha_1(\text{if A is present}) + \alpha_2(\text{if B is present}) + \alpha_3(\text{if C is present}) + \alpha_4(\text{if D is present}) + \alpha_5(\text{if E is present})$$

where P is the predicted probability of having an NSAID prescription and $\alpha_0 + \alpha_1 + \alpha_2 + \alpha_3 + \alpha_4 + \alpha_5 = 1$. The predicted probability increases as the number of codes increases. One assumption of regression is that the 5 codes are independent (and uncorrelated). For example, that would mean that if A=diabetes and B=congestive heart failure (CHF) then the likelihood of someone with diabetes having CHF is no greater than the likelihood of someone without diabetes having CHF. However, since diabetes can lead to heart disease, this assumption of independence is clearly false.

In our example, there are three variables used, each one giving a diagnosis for the patient, with the diagnosis related to the prescription value. In order to use logistic regression, we go to Analyze>Regression>Logistic. The next step is to assign tasks. We want to compare the hypothesis H_0: no relationship between the prescriptions and the three diagnosis codes versus H_1: some relationship exists.

Display 5.31. Screen for Logistic Regression

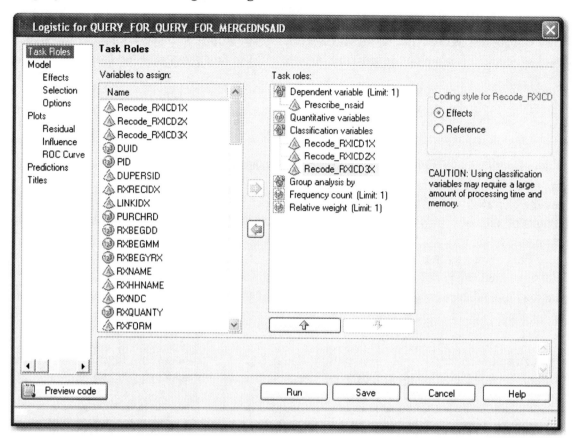

We next go to the Model Effects screen. The variables listed on the right need to be highlighted. Then the "Main" button is used to move the variables to the right hand side. Once in the window on the right, the variables are used in the model. For now, we will just look at the Plot>ROC value. The ROC curve will only work if Y has just two levels.

Display 5.32. Model Screen for Logistic Regression

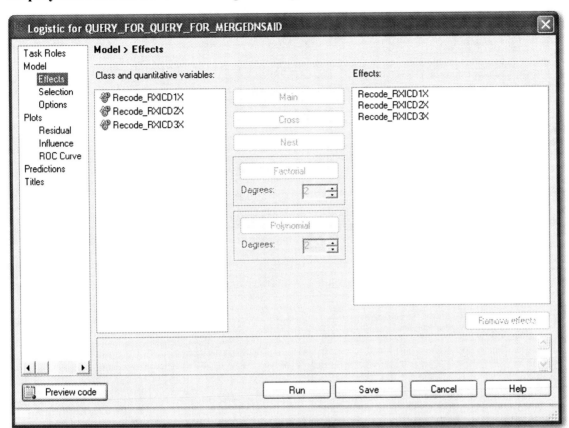

Display 5.33. Screen for Plots

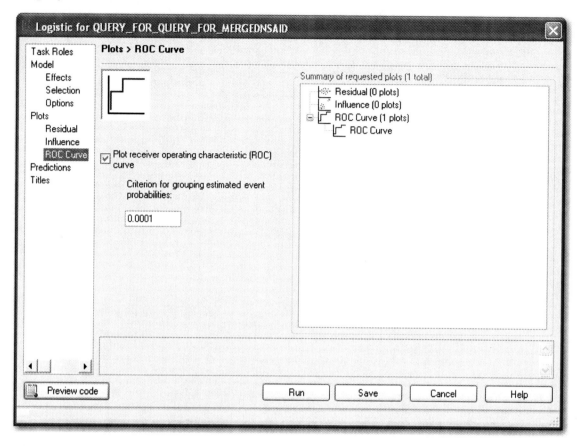

There are several parts to the logistic regression output.

Table 5.9. Summary Table for Logistic Regression

Model Information	
Data Set	WORK.SORTTEMPTABLESORTED
Response Variable	Prescribe_nsaid
Number of Response Levels	2
Model	binary logit
Optimization Technique	Fisher's scoring

The first table gives a summary, giving the Y variable and a description of the model.

Table 5.10. Number of Observations Used in the Model

Number of Observations Read	19572
Number of Observations Used	19572

Table 5.11. Y Response for Each Observation

Response Profile		
Ordered Value	Prescribe_nsaid	Total Frequency
1	0	19159
2	1	413

The next two tables give a summary of the number of observations. The discrepancy in group sizes is considerable, and impact the outcome.

Probability modeled is Prescribe_nsaid='0'.

Table 5.12. Summary of X-Variables Defined as Class

Class Level Information		
Class	Value	Design Variables
Recode_RXICD1X	-1	1　　　0
	0	0　　　1
	1	-1　　-1
Recode_RXICD2X	-1	1　　　0
	0	0　　　1
	1	-1　　-1
Recode_RXICD3X	-1	1　　　0
	0	0　　　1
	1	-1　　-1

Because the variables were identified as class variables, a summary of each one is also provided.

Model Convergence Status

Convergence criterion (GCONV=1E-8) satisfied.

Because a logistic regression model exists under certain criteria, the validity becomes questionable if it does not converge.

Table 5.13. Model Fit Statistics

Criterion	Intercept Only	Intercept and Covariates
AIC	4006.268	3288.827
SC	4014.150	3344.000
-2 Log L	4004.268	3274.827

For now, we will not concern ourselves with the model fit statistics. It is beyond the scope of this course.

Table 5.14. Chi-Square Test of Hypothesis

Testing Global Null Hypothesis: BETA=0			
Test	Chi-Square	DF	Pr > ChiSq
Likelihood Ratio	729.4406	6	<.0001
Score	1053.4259	6	<.0001
Wald	205.8534	6	<.0001

Table 5.15. Chi Square Test for Each X-Variable

Type 3 Analysis of Effects			
Effect	DF	Wald Chi-Square	Pr > ChiSq
Recode_RXICD1X	2	200.7984	<.0001
Recode_RXICD2X	2	175.3552	<.0001
Recode_RXICD3X	2	0.0044	0.9978

These chi-square statistics are important and give the probability of being wrong when rejecting H_0. In this problem, we can reject H_0. In addition, the first two diagnosis codes are statistically significant; the third code is not.

Table 5.16. Equation Coefficients

Analysis of Maximum Likelihood Estimates						
Parameter		DF	Estimate	Standard Error	Wald Chi-Square	Pr > ChiSq
Intercept		1	20.0788	254.9	0.0062	0.9372
Recode_RXICD1X	-1	1	2.2409	0.1799	155.1630	<.0001
Recode_RXICD1X	0	1	-4.7912	0.3427	195.4155	<.0001
Recode_RXICD2X	-1	1	2.8047	0.3303	72.0897	<.0001
Recode_RXICD2X	0	1	-5.0801	0.3847	174.3476	<.0001
Recode_RXICD3X	-1	1	6.4539	318.6	0.0004	0.9838
Recode_RXICD3X	0	1	-13.3435	254.9	0.0027	0.9583

Again, we will not concern ourselves with the above table.

Table 5.17. Odds Ratios

Odds Ratio Estimates			
Effect	Point Estimate	95% Wald Confidence Limits	
Recode_RXICD1X -1 vs 1	0.734	0.583	0.924
Recode_RXICD1X 0 vs 1	<0.001	<0.001	0.002
Recode_RXICD2X -1 vs 1	1.698	0.592	4.869
Recode_RXICD2X 0 vs 1	<0.001	<0.001	0.002
Recode_RXICD3X -1 vs 1	0.647	<0.001	>999.999
Recode_RXICD3X 0 vs 1	<0.001	<0.001	>999.999

The odds ratios are of interest. Those that are statistically significant indicate that there is a difference between items. However, the items are only significant if the confidence limits do not cross the value, one. If one is in the interval, then the values are not statistically significant. In this particular problem, the first two, comparing 1 versus -1 and 0 versus 1 are significant in the first diagnosis code; the comparison of 0 versus 1 in the second code is significant as well. What it means is that as the value of RXICD1X increases from -1 and 1 or from 0 to 1, the value of Y is likely to decrease from 0 to 1 (since the value of the odds ratio is less than 1). If the odds ratio is greater than one, then the value of Y increases as the value of the X-variable increases.

Table 5.18. Accuracy of Prediction

Association of Predicted Probabilities and Observed Responses			
Percent Concordant	48.5	Somers' D	0.320
Percent Discordant	16.5	Gamma	0.492
Percent Tied	35.0	Tau-a	0.013
Pairs	7912667	c	0.660

Table 5.17. gives the level of accuracy of the model. It indicates that the model predicts the value of Y correctly in fewer than half of the observations. The c-statistic (0.660) also is a measure of the accuracy of the model. It gives the area under the curve pictured below.

The terms specificity and sensitivity are defined as follows (Table 5.18).

Table 5.19. Table of Sensitivity and Specificity

		Outcome Value (Y)		
		PRESENT	ABSENT	
Model value	+	a	b	a + b
	-	c	d	c + d
		a + c	b + d	a + b + c + d

Then

SENSITIVITY = a/(a+c)
SPECIFICITY = d/(b+d).

In other words, specificity is the probability of the model predicting 0 when the Y value is in fact 0; sensitivity is the probability predicting that Y is one given that Y is in fact one. The ROC curve graphs sensitivity against 1-specificity. The closer the area under the curve is to one, the more accurate the model (Figure 5.2). The c-statistic in Table 5.17 is equal to 0.660, the measure of the area under the ROC curve.

Figure 5.2. Graph of ROC Curve

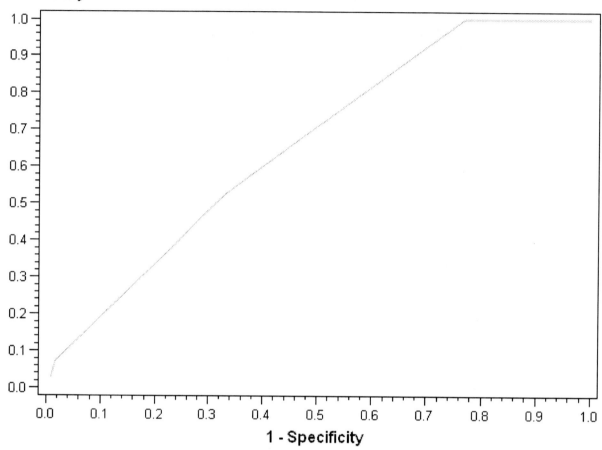

Logistic regression works best when the group sizes are approximately the same size. Therefore, we keep the NSAID observations, and take a sample of the non-NSAID drugs. If this is done, the accuracy is equal to the values in Table 5.20.

Table 5.20. Accuracy of Model

Actual Value	Predicted Value	Predictive accuracy
0	0	100%
1	0	77%
1	1	23%

Overall, the misclassification rate is 38%.

However, what these tests do is to compare the groups of heart failure, arthritis, and neither. This actually gets away from our initial objective of wanting to determine whether NSAIDs are prescribed to patients with heart failure, and what type. We look at the situation for heart failure. In this case, we look at the variable, RXStrength since these patients should be prescribed low-dose aspirin. The following table gives the different strengths prescribed.

Table 5.21. Strengths of Medications

Strength	Frequency	Percent
325	72	58
5	14	11
50	9	7
600	4	3
81	24	20

Only 20% are prescribed the low-dose (81 mg). The more common value is standard aspirin at 325. The 5, 50, and 600 are non-aspirin NSAID.

5.6. Exercises

1. Use your hypotheses in Chapter 4, and analyze them using chi square and logistic regression.

2. Use your hypotheses concerning the MEPS data, and analyze them using chi square and logistic regression.

Chapter 6. Linear Regression (Analysis of Variance)

6.1. Introduction 215

6.2. Linear Models in Enterprise Guide 216

6.3. Exercises 248

6.1. Introduction

The general linear model is used to perform linear regression. Linear regression assumes that an interval (continuous) outcome variable is a linear relationship of a series of input variables. These input variables can be nominal, ordinal, or interval. For example, suppose we want to examine the relationship of teaching time to research time in the workload database.

The methodology looks very similar to that of logistic regression in that

$$Y = \beta_0 + \beta_1 X_1 + \beta_2 X_2 + \ldots + \beta_n X_n.$$

However, in this case, Y is assumed to be an interval, or continuous, variable. Again, $X_1,\ldots,X_n$ can be continuous, ordinal, or nominal.

We also must assume that there is an error term, ε_i where ε_i is the difference between the actual value of Y and the predicted value of Y. Then we assume that the values of $\varepsilon_1,\varepsilon_2,\ldots,\varepsilon_n$ are from a normal distribution (with a bell-shaped density curve). Moreover, the average value of the $\varepsilon_1,\varepsilon_2,\ldots,\varepsilon_n$ must equal zero. Moreover, if ε_i is graphed against Y_i, the graph must look randomly scattered. If it does not, the basic assumptions of the model are not satisfied. These assumptions should be checked whenever you do a linear analysis.

6.2. Linear Models in Enterprise Guide

In Enterprise Guide, we use linear models with the starting point shown in the following menu (Display 6.1).

Display 6.1. Menu for Linear Models

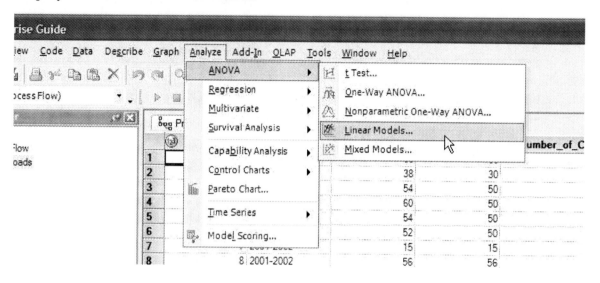

Then we identify the tasks (Display 6.2).

Display 6.2. Screen for Assigning Values

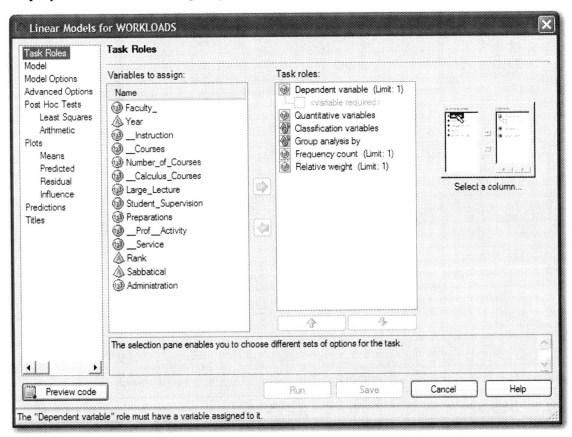

In this example, we use instruction as the input variable and research as the output variable from the Workloads dataset. Y=dependent variable. X=class variable if it is nominal or ordinal; otherwise, X=quantitative variable.

Display 6.3. Assignment of X and Y Variables

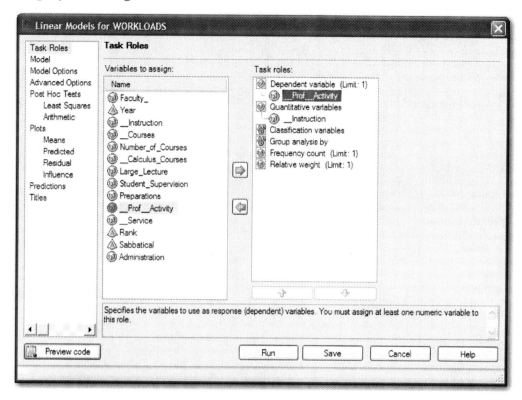

Next, we have to identify the model. In this example, we need to define instruction as a main effect (Display 6.4).

Display 6.4. Entering X-Variables Into Model

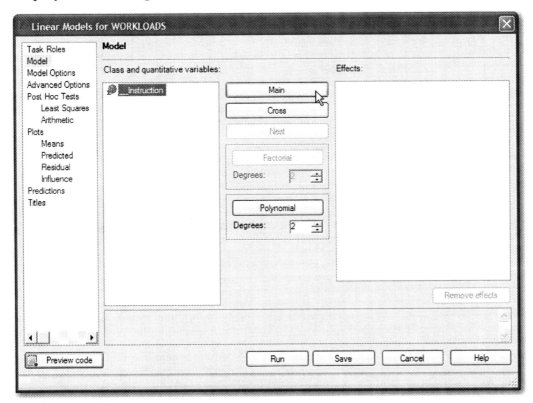

We also go to plots for the predicted values, and for the residuals (Display 6.5).

Display 6.5. Check Residual Plots to Verify Assumptions

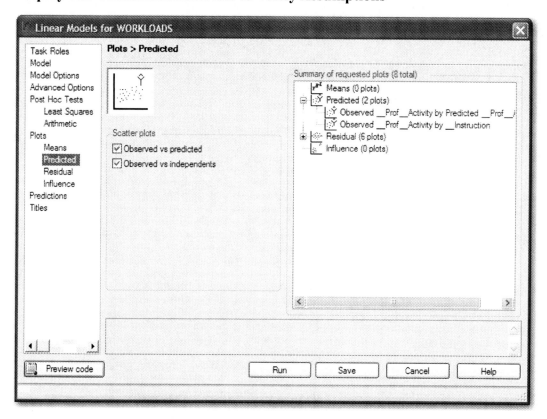

Display 6.6. Residuals, Continued

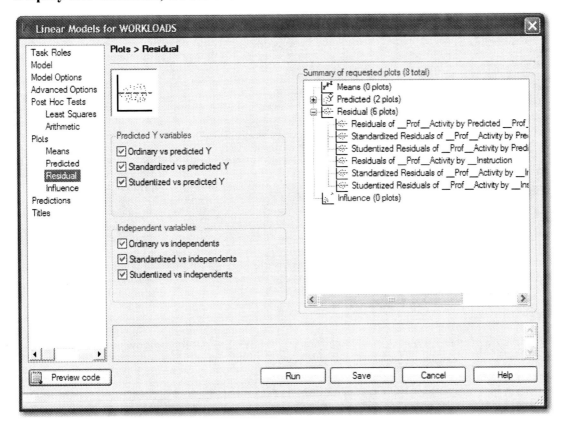

The results are given below (Tables 6.1-6.6).

Table 6.1. Number of Observations Used in Model

The GLM Procedure

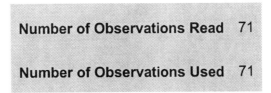

First, the result gives the number of observations in the dataset that do not contain missing values in either the input or outcome variables.

Table 6.2. Overall Model Results

Dependent Variable: __Prof__Activity % Prof# Activity

Source	DF	Sum of Squares	Mean Square	F Value	Pr > F
Model	1	13340.34840	13340.34840	56.67	<.0001
Error	69	16244.24315	235.42381		
Corrected Total	70	29584.59155			

This box gives the overall results of the model. The p-value is statistically significant (<0.0001) meaning that the input variable(s) explain the variability in the outcome variable.

Table 6.3. R-Square Value

R-Square	Coeff Var	Root MSE	__Prof__Activity Mean
0.450922	39.48497	15.34353	38.85915

The r-square value gives the percentage of variability in the outcome variable that can be explained by the input variable(s). For this example, 45% of the variability in research is explained by the variability in instruction. Thus, 55% of the variability in Y (professional activity) remains unexplained.

Table 6.4. Type I Sum of Squares

Source	DF	Type I SS	Mean Square	F Value	Pr > F
__Instruction	1	13340.34840	13340.34840	56.67	<.0001

Table 6.5. Type III Sum of Squares

Source	DF	Type III SS	Mean Square	F Value	Pr > F
__Instruction	1	13340.34840	13340.34840	56.67	<.0001

Table 6.6. Coefficients of the Model

Parameter	Estimate	Standard Error	t Value	Pr > |t|
Intercept	72.80420800	4.86317335	14.97	<.0001
Instruction	-0.81587636	0.10838420	-7.53	<.0001

Table 6.4 gives the Type I SS, Table 6.5 gives the Type III SS, and Table 6.6 gives an estimate of the coefficient of the x-variables. In this example, as there is only one x-variable, the three tables provide exactly the same information. Generally, it is better to use the Type III SS because they are order-invariant, meaning that the values don't change if the order of the x-variables changes. In this example, the tables again show that instruction is a statistically significant predictor of the time spent in research. Figure 6.1 gives a graph of the actual professional activity percentage versus the predicted percentage. The is followed by the actual predictions (versus actual results). While the graph is generally linear, there is scattering, indicating that some values are more difficult to predict compared to others.

Figure 6.1. Graph of Predicted Value Versus Actual Value

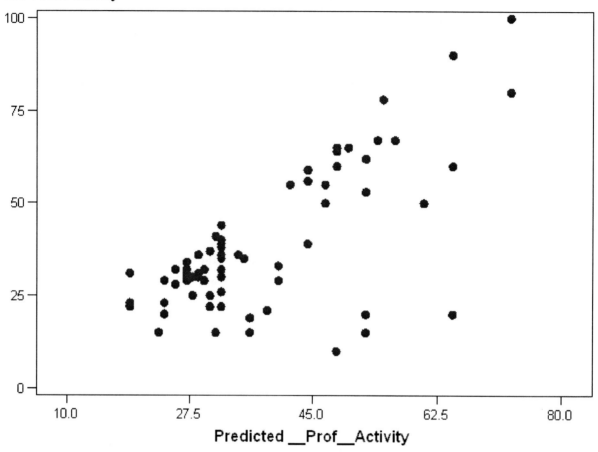

We should expect this graph to look somewhat linear. The closer the value of R^2 is to 1, the more linear this graph will appear. Table 6.7 gives a listing of the actual and predicted values.

Table 6.7. Predicted and Actual Values

Predicted __Prof__Activity	% Prof# Activity
18.9563682261959	23
41.8009063129872	55
28.7468845491064	31
23.8516263876511	20
28.7468845491064	36
30.3786372695915	25

Predicted __Prof__Activity	% Prof# Activity
60.5660625985657	50
27.1151318286213	32
18.9563682261959	31
52.4072989961403	20
27.1151318286213	31
18.9563682261959	22
32.0103899900766	39
64.6454443997784	60
56.486680797353	67
23.0357500274086	15
31.1945136298341	15
44.2485353937148	39
23.8516263876511	29
27.9310081888639	30
32.0103899900766	22
27.1151318286213	34
27.1151318286213	30
32.0103899900766	26
32.0103899900766	38
48.3279171949275	60
29.562760909349	29
25.4833791081362	28
25.4833791081362	32

Predicted __Prof__Activity	% Prof# Activity
30.3786372695915	37
64.6454443997784	20
52.4072989961403	15
44.2485353937148	56
38.537400872017	21
32.0103899900766	36
29.562760909349	32
54.0390517166254	67
64.6454443997784	90
52.4072989961403	53
46.6961644744424	55
72.8042080022039	100
40.1691535925021	29
44.2485353937148	59
72.8042080022039	100
32.0103899900766	30
32.0103899900766	35
23.8516263876511	23
27.9310081888639	25
40.1691535925021	33
32.0103899900766	44
32.0103899900766	32
25.4833791081362	32

Predicted __Prof__ Activity	% Prof# Activity
28.7468845491064	30
72.8042080022039	80
48.3279171949275	10
52.4072989961403	15
31.1945136298341	41
49.9596699154126	65
34.4580190708042	36
48.3279171949275	65
54.8549280768679	78
36.0897717912893	15
48.3279171949275	10
52.4072989961403	62
48.3279171949275	64
27.1151318286213	29
35.2738954310468	35
36.0897717912893	19
32.0103899900766	40
30.3786372695915	22
46.6961644744424	50

Figure 6.2. X Versus Y Variable

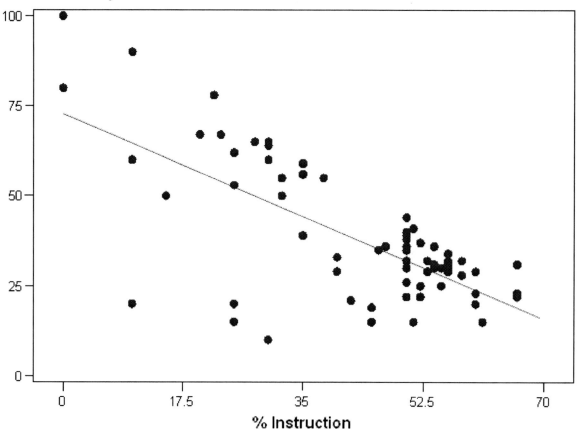

This graph compared the value of the percent of time in research compared to the percent of time in instruction. The curve is generally downward. However, it shows that the scattering is fan-shaped. That is, the lower the level of instruction, the harder it is to predict the value of research. For this reason, the assumption of equal variance is violated, and the adequacy of the model comes into question.

Figure 6.3. Y Versus Residuals

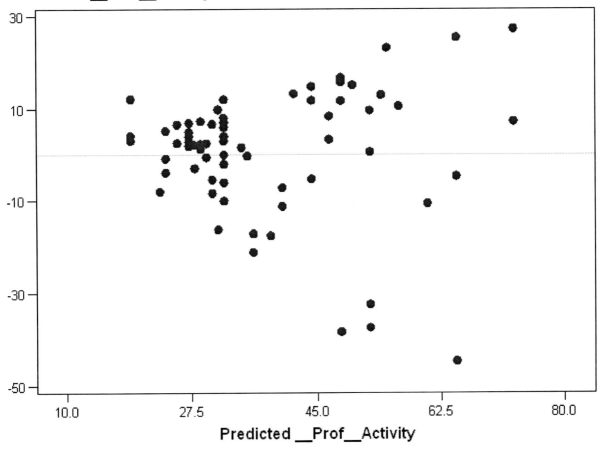

Figure 6.3 is fan-shaped rather than random. It indicates that the basic model assumptions are not valid. While there are methods to adjust the model to make the assumptions valid, they are beyond the scope of this course. **Table 6.8. Values of Professional Activity Versus Residuals**

Predicted __Prof__Activity	Residual of __Prof__Activity
18.9563682261959	4.04363177380408
41.8009063129872	13.1990936870128
28.7468845491064	2.25311545089353
23.8516263876511	-3.85162638765119
28.7468845491064	7.25311545089353
30.3786372695915	-5.37863726959156

Predicted __Prof__Activity	Residual of __Prof__Activity
60.5660625985657	-10.5660625985657
27.1151318286213	4.88486817137862
18.9563682261959	12.043631773804
52.4072989961403	-32.4072989961403
27.1151318286213	3.88486817137862
18.9563682261959	3.04363177380408
32.0103899900766	6.98961000992334
64.6454443997784	-4.64544439977848
56.486680797353	10.5133192026469
23.0357500274086	-8.03575002740864
31.1945136298341	-16.1945136298341
44.2485353937148	-5.24853539371483
23.8516263876511	5.1483736123488
27.9310081888639	2.06899181113607
32.0103899900766	-10.0103899900766
27.1151318286213	6.88486817137862
27.1151318286213	2.88486817137862
32.0103899900766	-6.01038999007665
32.0103899900766	5.98961000992334
48.3279171949275	11.6720828050724
29.562760909349	-0.56276090934901
25.4833791081362	2.51662089186371
25.4833791081362	6.51662089186371

Predicted __Prof__Activity	Residual of __Prof__Activity
30.3786372695915	6.62136273040844
64.6454443997784	-44.6454443997784
52.4072989961403	-37.4072989961403
44.2485353937148	11.7514646062851
38.537400872017	-17.537400872017
32.0103899900766	3.98961000992334
29.562760909349	2.43723909065098
54.0390517166254	12.9609482833746
64.6454443997784	25.3545556002215
52.4072989961403	0.5927010038597
46.6961644744424	8.30383552555751
72.8042080022039	27.195791997796
40.1691535925021	-11.1691535925021
44.2485353937148	14.7514646062851
72.8042080022039	27.195791997796
32.0103899900766	-2.01038999007665
32.0103899900766	2.98961000992334
23.8516263876511	-0.85162638765119
27.9310081888639	-2.93100818886392
40.1691535925021	-7.16915359250211
32.0103899900766	11.9896100099233
32.0103899900766	-0.01038999007665
25.4833791081362	6.51662089186371

Predicted __Prof__Activity	Residual of __Prof__Activity
28.7468845491064	1.25311545089353
72.8042080022039	7.19579199779605
48.3279171949275	-38.3279171949275
52.4072989961403	-37.4072989961403
31.1945136298341	9.80548637016589
49.9596699154126	15.0403300845873
34.4580190708042	1.5419809291957
48.3279171949275	16.6720828050724
54.8549280768679	23.145071923132
36.0897717912893	-21.0897717912893
48.3279171949275	-38.3279171949275
52.4072989961403	9.5927010038597
48.3279171949275	15.6720828050724
27.1151318286213	1.88486817137862
35.2738954310468	-0.27389543104683
36.0897717912893	-17.0897717912893
32.0103899900766	7.98961000992334
30.3786372695915	-8.37863726959156
46.6961644744424	3.30383552555751

Observation of Table 6.8 shows that for large values of professional activity, there are large values for residuals and conversely. The values do not occur randomly.

Standardization is not necessary if only one x-variable is used. The purpose of standardization is to ensure that a variable does not dominate the rest simply because its scale is much larger compared to the scale of other variables.

Since standardization is not important here, we will omit an examination of the output. We will run the analysis again, adding two additional x-variables into the model. In this example, we will include both rank and year.

Display 6.7. Additional X-Variables

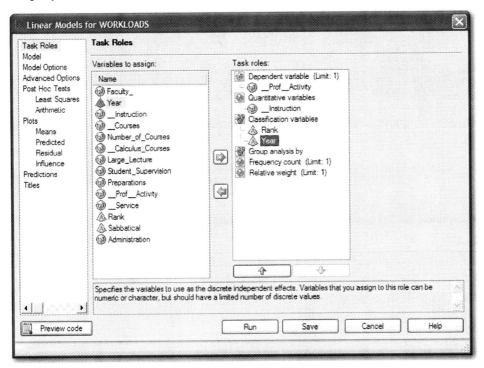

The two new variables are class variables rather than interval. We will also add an interaction to the model to determine whether rank interacts with year (Display 6.8).

Display 6.8. Defining Model With Interaction

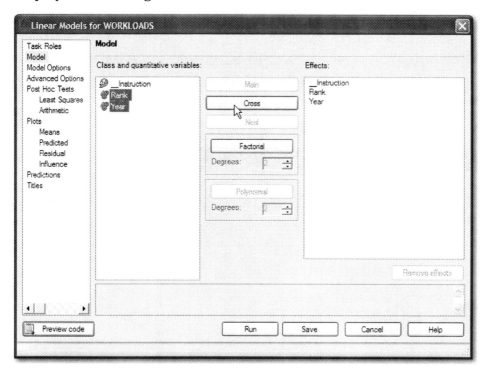

Highlight the two variables and then press the cross button. By using the Cross button, the interaction is added to the model (Display 6.9).

Display 6.9. Addition of Crossed Variable

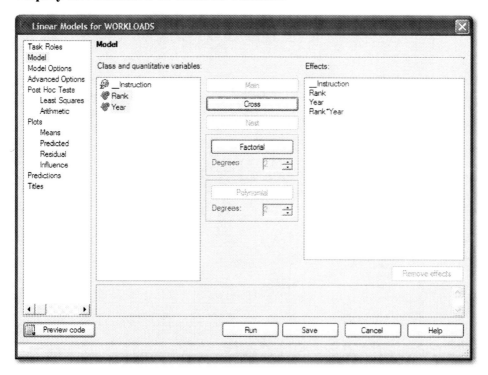

Because we are using classification variables, we can now include post-hoc tests. These should only be used if there is statistical significance in the overall model. The least squares tests should be used if there are multiple x-variables. The least squares tests isolate the impact of one variable while accounting for the impact of the other variables.

Display 6.10. Definition of Post-Hoc Tests

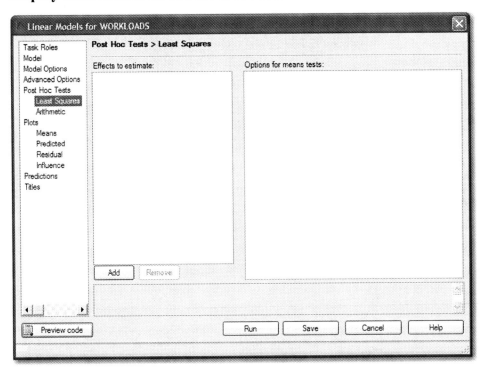

Use the add button. Then turn the values from False to True on the right side window (Display 6.11).

Display 6.11. Development of Post-Hoc Tests

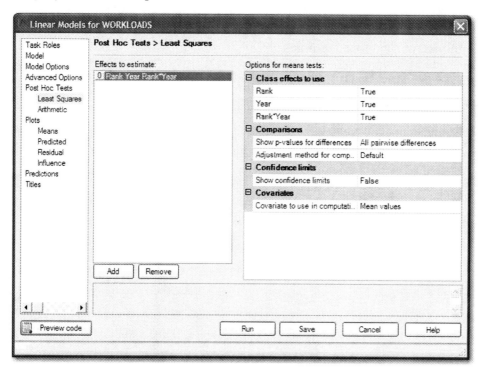

The results are given below in Tables 6.9-6.15.

Table 6.9. Definition of Class Variables

The GLM Procedure

Class Level Information

Class	Levels	Values
Rank	3	Assistant Associate Professor
Year	3	2001-2002 2002-2003 2003-2004

Table 6.10. Number of Observations Used in the Model

Number of Observations Read	71
Number of Observations Used	71

Note that with class variables, all possible levels are given.

Table 6.11. Overall Model Results

Dependent Variable: __Prof__Activity % Prof# Activity

Source	DF	Sum of Squares	Mean Square	F Value	Pr > F
Model	9	16033.38698	1781.48744	8.02	<.0001
Error	61	13551.20456	222.15089		
Corrected Total	70	29584.59155			

The overall model is statistically significant.

Table 6.12. R-Square Value

R-Square	Coeff Var	Root MSE	__Prof__Activity Mean
0.541951	38.35577	14.90473	38.85915

Adding the new variables increased the value of the R-square from 45% to 54%, indicating that 46% of the variability in research levels is still not accounted for.

Table 6.13. Type I Sum of Squares

Source	DF	Type I SS	Mean Square	F Value	Pr > F
Instruction	1	13340.34840	13340.34840	60.05	<.0001
Rank	2	2510.41212	1255.20606	5.65	0.0056
Year	2	60.72236	30.36118	0.14	0.8725
Rank*Year	4	121.90411	30.47603	0.14	0.9679

Table 6.14. Type III Sum of Squares

Source	DF	Type III SS	Mean Square	F Value	Pr > F
Instruction	1	13118.52956	13118.52956	59.05	<.0001
Rank	2	2303.34751	1151.67375	5.18	0.0083
Year	2	56.98559	28.49279	0.13	0.8799
Rank*Year	4	121.90411	30.47603	0.14	0.9679

Now that there is more than one variable, the Type I SS and the Type III SS are different. Ignore the Type I SS and use the Type III SS. The Type III SS indicate that instruction and rank are statistically significant. Since Year and the interaction of Rank*Year are not significant, we can ignore any additional analysis of those variables. The table below gives the coefficients of the regression equation.

Table 6.15. Coefficients of Equation

Parameter	Estimate		Standard Error	t Value	Pr > |t|
Intercept	71.10985795	B	6.12756143	11.60	<.0001
Instruction	-0.99093176		0.12895115	-7.68	<.0001
Rank Assistant	16.92766191	B	8.88541471	1.91	0.0615
Rank Associate	13.09106833	B	7.14624099	1.83	0.0719
Rank Professor	0.00000000	B	.	.	.
Year 2001-2002	0.34247084	B	6.57770989	0.05	0.9586
Year 2002-2003	4.43623139	B	6.42600120	0.69	0.4926
Year 2003-2004	0.00000000	B	.	.	.
Rank*Year Assistant 2001-2002	-2.85153908	B	13.14543861	-0.22	0.8290
Rank*Year Assistant 2002-2003	-2.56381017	B	12.36044059	-0.21	0.8364
Rank*Year Assistant 2003-2004	0.00000000	B	.	.	.
Rank*Year Associate 2001-2002	1.60713567	B	9.83389230	0.16	0.8707
Rank*Year Associate 2002-2003	-4.35973756	B	9.84413211	-0.44	0.6594
Rank*Year Associate 2003-2004	0.00000000	B	.	.	.
Rank*Year Professor 2001-2002	0.00000000	B	.	.	.
Rank*Year Professor 2002-2003	0.00000000	B	.	.	.
Rank*Year Professor 2003-2004	0.00000000	B	.	.	.

Each level of a class variable gets a coefficient.

Note: The X'X matrix has been found to be singular, and a generalized inverse was used to solve the normal equations. Terms whose estimates are followed by the letter 'B' are not uniquely estimable.

Generated by the SAS System (Local, XP_PRO) on 28MAY2007 at 2:39 PM

Note that we may now have a singular matrix. Since Year and the Interaction of Rank*Year are not significant, we may safely remove those variables from the model to see if we remove the singularity.

Display 6.12. Model Modification

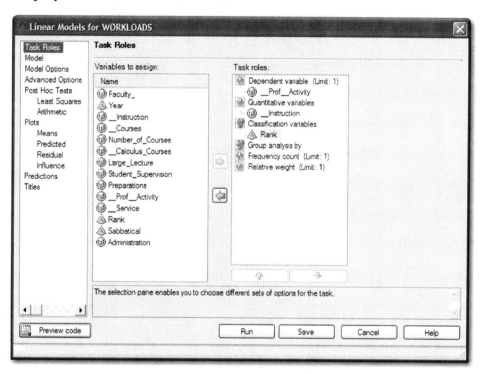

The GLM Procedure

Class Level Information		
Class	Levels	Values
Rank	3	Assistant Associate Professor

Number of Observations Read	71
Number of Observations Used	71

Dependent Variable: __Prof__Activity % Prof# Activity

Source	DF	Sum of Squares	Mean Square	F Value	Pr > F
Model	3	15850.76052	5283.58684	25.78	<.0001
Error	67	13733.83103	204.98255		
Corrected Total	70	29584.59155			

R-Square	Coeff Var	Root MSE	__Prof__Activity Mean
0.535778	36.84386	14.31721	38.85915

When removing these two variables, the R-squared value changes only marginally; they are not necessary to the equation.

Source	DF	Type I SS	Mean Square	F Value	Pr > F
__Instruction	1	13340.34840	13340.34840	65.08	<.0001
Rank	2	2510.41212	1255.20606	6.12	0.0036

Source	DF	Type III SS	Mean Square	F Value	Pr > F
__Instruction	1	15426.43290	15426.43290	75.26	<.0001
Rank	2	2510.41212	1255.20606	6.12	0.0036

Parameter	Estimate		Standard Error	t Value	Pr > |t|
Intercept	72.91121400	B	4.54624703	16.04	<.0001
__Instruction	-0.99925434		0.11518654	-8.68	<.0001
Rank Assistant	15.50876017	B	5.27062155	2.94	0.0045
Rank Associate	12.53464755	B	4.18157748	3.00	0.0038
Rank Professor	0.00000000	B	.	.	.

Note: The X'X matrix has been found to be singular, and a generalized inverse was used to solve the normal equations. Terms whose estimates are followed by the letter 'B' are not uniquely estimable.

Generated by the SAS System (Local, XP_PRO) on 28MAY2007 at 2:56 PM

Unfortunately, we did not remove the singularity. Nevertheless, we will continue the analysis.

Table 6.16. Post-Hoc Comparisons

Adjustment for Multiple Comparisons: Tukey-Kramer

Rank	__Prof__Activity LSMEAN	LSMEAN Number
Assistant	46.8453638	1
Associate	43.8712512	2
Professor	31.3366037	3

The class levels are given coded values of 1,2,3.

Table 6.17. P-Values for Pairwise Comparisons

Least Squares Means for effect Rank
Pr > |t| for H0: LSMean(i)=LSMean(j)
Dependent Variable: __Prof__Activity

i/j	1	2	3
1		0.8285	0.0123
2	0.8285		0.0105
3	0.0123	0.0105	

The pairwise analysis suggests that professors differ from both assistant and associate professors in the percentage of research; however, assistant and associate professors do not show a statistically significant difference. Recall that if there is insufficient evidence to show a difference; that does not necessarily mean that a difference does not exist.

The relationship between the actual and predicted values looks slightly more linear compared to the first analysis. This reflects the improvement in the R-square value (Figure 6.4).

Figure 6.4. Relationship Between Actual and Predicted Values

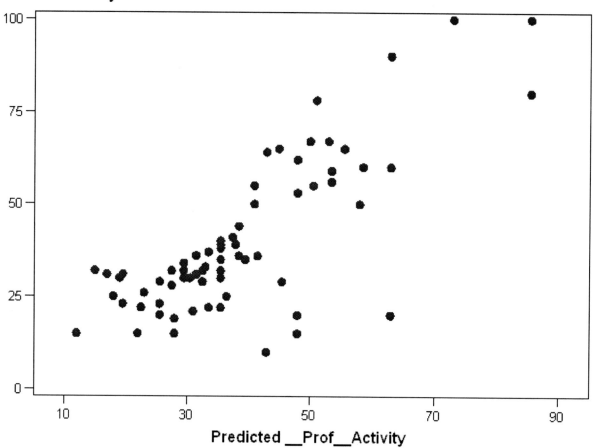

However, the relationship between professional activity and instruction is still a fan-shape, indicating that the prediction is more effective a higher values of instruction (Figure 6.5).

Figure 6.5. Instruction Versus Professional Activity

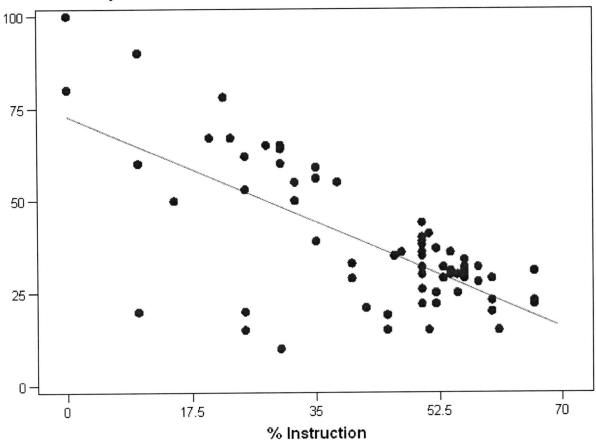

Figure 6.6. Y Versus Class X.

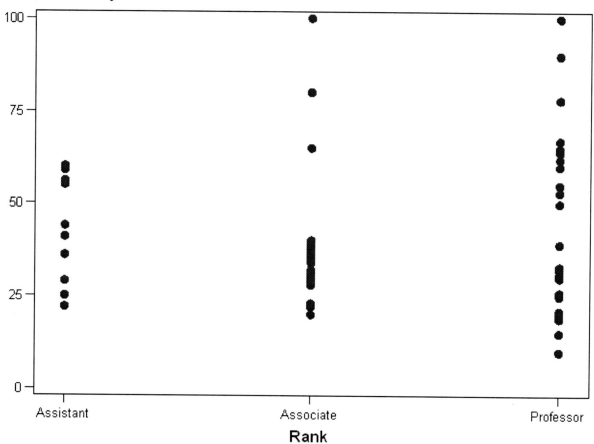

Because rank is a class variable, the relationship will look like this. It is better to examine the average level of activity for each rank.

Figure 6.7. Residuals Versus Predictions

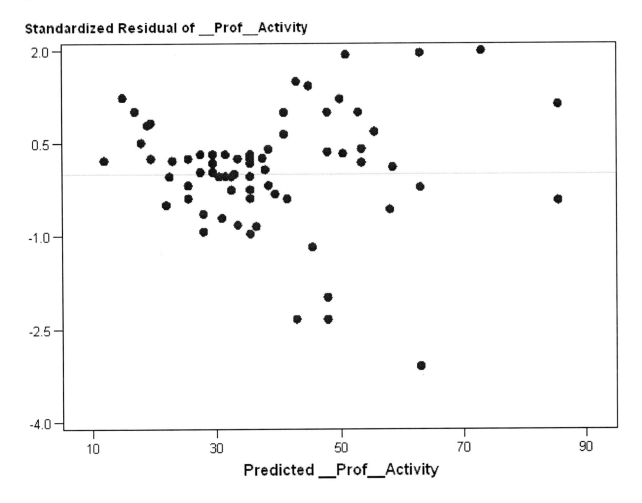

The residual plot looks more scattered than fan-shaped, indicating that the assumption of equal variances is likely valid.

As a conclusion, there is a relationship as defined by the model. However, the assumptions of the model are of somewhat questionable validity.

6.3. Exercises

1. Use the MEPS database, and investigate the relationship of self-pay and private-pay to total pay.

2. Use the survey data with hours as the Y variable.

Chapter 7. Additional Examples from Student Papers

7.1. Example 1 of Logistic Regression 251

7.2. Example 2-Linear Models 259

7.1. Example 1 of Logistic Regression

An Analysis of Barry Bonds' Record Breaking 2001 Season

Abstract

The objective of this paper is to analyze statistics collected from Major League Baseball All-Star Barry Bonds' record breaking 2001 season. This will be accomplished by organizing the strongly categorical dataset into various tables that will provide insight into different game situations and how Barry Bonds handles them. In addition to table analyses, logistic regression will be performed to see whether or not different variables were a significant predictor of his performance at the plate. A great deal of planning and strategic maneuvering goes on during every Major League game and some of this will become apparent when details of Bonds' situational performance are explored.

Results from the various analyses that took place uncovered many of the intricacies of the game of baseball as it is played in the modern era. Different game situations ended up altering Bonds' performance quite a bit in several areas as would be expected not only for Bonds but for every Major League Baseball player, including his performance at the beginning and end of games and with a different number of outs left in innings.

Introduction

In 2001, Barry Bonds, an outfielder for the San Francisco Giants, broke the Major League Baseball single season home run record with 71 home runs. During the same season, he also led the Major Leagues in several other categories, including walks and slugging percentage. Bonds' success in 2001 led to statisticians compiling detailed records of each and every plate appearance and what occurred during them. The dataset being analyzed here is precisely that, and consists of 15 variables, mostly categorical in nature, which describe the conditions and results from all 648 at-bats that Barry Bonds had during the 2001 MLB season.

The abundance of categorical variables in the dataset lends itself to many opportunities for interesting table analyses that will more closely examine how Barry Bonds performed in various situations. In addition, a logistic regression analysis will be performed that explores his performance patterns over the continuous variable of time.

Results showed a tendency for Bonds to perform the best during the first half of games and with fewer outs left in innings. In addition, opposing teams walked him a great deal at the beginning and end of games for strategic reasons. Patterns relating the variability in his performance to the variability in different game situations were also discovered.

Method

The dataset used in this analysis was obtained from the Journal of Statistics Education Data Archive (http://www.amstat.org/publications/jse/jse_data_archive.html) after being submitted by Jerome P. Reiter of the Institute of Statistics and Decision Sciences at Duke University. The data are complete with the small exception of some pitcher data that was missing at random and was not used in the current analysis.

Once the data were organized into a .dat file on the website, it was downloaded and converted into a SAS dataset for analysis. The variables were all numeric in nature and all but two of them were categorical. The non-categorical variables were the plate appearances counter and the opposing pitchers earned-run average (ERA).

Analysis of this strongly categorical dataset centered around tables and examining how performance changed in different situations.

Data Analysis

One of the most important and basic things to analyze when it comes to a baseball players' performance is what exactly he did when he was up to bat. There was a column in the dataset labeled "Result" consisting of integer values between 0 and 5 that represented the six possible outcomes of a single at-bat for Barry Bonds. The outcomes and their corresponding integer codes were as follows:

> 0 = Bonds did not reach base
>
> 1 = Bonds reached first base on a single or error
>
> 2 = Bonds reached second base on a double or error
>
> 3 = Bonds reached third base on a triple or error
>
> 4 = Bonds hit a home run
>
> 5 = Bonds walked or was hit by a pitch

In order to get a summary of how often each of these outcomes occurred, a one-way frequency analysis was performed on the dataset. The results are shown in Table 1.

Table 1. Barry Bonds' Summary of Performance

One-Way Frequencies
Results

The FREQ Procedure

Result				
Result	Frequency	Percent	Cumulative Frequency	Cumulative Percent
0	315	48.61	315	48.61
1	49	7.56	364	56.17
2	33	5.09	397	61.27
3	2	0.31	399	61.57
4	71	10.96	470	72.53
5	178	27.47	648	100.00

Table 1 clearly shows the record-breaking 71 home runs (Result = 4) hit by Barry Bonds in the 2001 season, which amounted to an astonishing 10.96% of his plate appearances. The frequencies of the other results were similar to what was expected. Bonds walked or was hit by a pitch 178 times (Result = 5), which is very high amongst major league hitters and can be attributed to opposing pitchers and teams strategically avoiding giving Bonds anything to hit in close games because of his offensive potency.

Next it will be necessary to do some Table Analysis to see how well Bonds does in different situations. First, the Result column will be broken down by inning to see if there are any performance differences at the beginning and end of games. Major League Baseball games are scheduled for 9 innings with at least 3 batters making an appearance in each inning, which generally results in each batter stepping up to the plate between 3 and 5 times per game. Table 2 shows Bonds' performance inning by inning.

Table 2. Barry Bonds Performance By Inning

The FREQ Procedure

Frequency Row Pct Col Pct	Table of Inning by Result						
	Result(Result)						
Inning(Inning)	0	1	2	3	4	5	Total
1	58 40.85 18.41	10 7.04 20.41	11 7.75 33.33	0 0.00 0.00	12 8.45 16.90	51 35.92 28.65	142
2	3 20.00 0.95	1 6.67 2.04	2 13.33 6.06	0 0.00 0.00	4 26.67 5.63	5 33.33 2.81	15
3	46 52.27 14.60	10 11.36 20.41	4 4.55 12.12	1 1.14 50.00	8 9.09 11.27	19 21.59 10.67	88
4	28 44.44 8.89	2 3.17 4.08	2 3.17 6.06	1 1.59 50.00	12 19.05 16.90	18 28.57 10.11	63
5	32 45.71 10.16	5 7.14 10.20	4 5.71 12.12	0 0.00 0.00	9 12.86 12.68	20 28.57 11.24	70
6	44 61.11 13.97	5 6.94 10.20	2 2.78 6.06	0 0.00 0.00	7 9.72 9.86	14 19.44 7.87	72
7	35 59.32 11.11	5 8.47 10.20	5 8.47 15.15	0 0.00 0.00	6 10.17 8.45	8 13.56 4.49	59
8	38 48.10 12.06	7 8.86 14.29	1 1.27 3.03	0 0.00 0.00	9 11.39 12.68	24 30.38 13.48	79
9	21 53.85 6.67	3 7.69 6.12	1 2.56 3.03	0 0.00 0.00	2 5.13 2.82	12 30.77 6.74	39

It is interesting to note the differences in Total appearances by inning. This is because Barry Bonds has batted third in the lineup for his team for most of his career, which means that he is guaranteed to make an appearance in the 1st inning and is unlikely to make an appearance in the 2nd inning unless his team is doing very well. From the 3rd inning on, the appearances are pretty steady.

When it comes to performance, Table 2 shows that the percentage for each result by inning is fairly consistent with perhaps a few exceptions. One that doesn't seem to be coincidence is that Bonds was walked or hit by a pitch (Result = 6) more at the very beginning and very end of games than he was in the middle. An explanation for this would be that at the beginning of games, pitchers are just getting into a rhythm themselves and they don't want good hitters to get into a rhythm by giving them hittable pitches. Then at the end of close games, intentionally walking dangerous batters is a smart and strategic move for a team so that they can avoid giving up big hits or home runs, especially in Bonds' case.

The next Table Analysis that will be done will show how Bonds performs when there are 0, 1 and 2 outs. Table 3 shows the results.

Table 3. Barry Bonds Performance By Number of Outs

Table Analysis
Results

The FREQ Procedure

Frequency Row Pct Col Pct	Table of NumOuts by Result						
	Result(Result)						
NumOuts(NumOuts)	0	1	2	3	4	5	Total
0	96 54.55 30.48	17 9.66 34.69	11 6.25 33.33	1 0.57 50.00	20 11.36 28.17	31 17.61 17.42	176
1	109 46.98 34.60	16 6.90 32.65	16 6.90 48.48	1 0.43 50.00	30 12.93 42.25	60 25.86 33.71	232
2	110 45.83 34.92	16 6.67 32.65	6 2.50 18.18	0 0.00 0.00	21 8.75 29.58	87 36.25 48.88	240
Total	315	49	33	2	71	178	648

It can be seen that the general trend for Results 1, 2, 3 and 4 is that they decrease as the number of outs increase. The translation is that Bonds tends to perform his best early in innings when there is hope of putting something together. Many players tend to look ahead to the next inning when there are already 2 outs and there isn't much scoring potential.

The other noticeable trend in this table is that Bonds is walked or hit by a pitch (Result = 6) much more as the number of outs goes up (18% to 26% to 36%). This makes perfect sense because with 0 or 1 out, opposing teams have little desire to put runners on base to potentially score when the inning has just begun, and with 2 outs, teams may decide to walk Bonds on purpose in certain situations since they only need 1 more out to end the inning.

A final analysis that will be done on this dataset is a logistic regression analysis of the dependent, categorical variable, Result (6 response levels), with the continuous variable, Appearance. Appearance has not been used yet and is simply a counter for each plate appearance by Barry Bonds (i.e. his first at-bat of the season has an Appearance value of '1', etc.). This analysis will check for consistency of performance throughout the course of the 2001 season. The initial summary results are listed in Figure 1.

Figure 1. Summary of Logistic Regression Analysis

Logistic Regression Results

The LOGISTIC Procedure

Model Information		
Data Set	WORK.SORTTEMPTABLESORTED	
Response Variable	Result	Result
Number of Response Levels	6	
Model	cumulative cloglog	
Optimization Technique	Fisher's scoring	

Number of Observations Read	648
Number of Observations Used	648

Response Profile		
Ordered Value	Result	Total Frequency
1	0	315
2	1	49
3	2	33
4	3	2
5	4	71
6	5	178

Above is a basic summary of the data in question. The 6 response levels are listed and correspond to the 6 possible outcomes in the Result column. Due to the large number of response levels, the Complementary Log-Link Function Model was used on the data. There were a total of 648 total observations, which means Barry Bonds went to the plate 648 times during the 2001 season. Figure 2 below shows the remainder of the results from the logistic regression analysis.

Figure 2. Detailed Logistic Regression Results

The above results show that the logistic regression did satisfy its convergence criterion with the chi-square values listed. The Analysis of Maximum Likelihood Estimates provides the coefficients of the regression equation and the Association of Predicted Probabilities and Observed Responses shows the level of accuracy of the model. This particular model predicted the outcome correctly in slightly more than half of the observations (53.9%). The c-statistic also implies a relatively low level of accuracy with a value of 0.546.

Discussion

This dataset provided a very interesting and detailed view into the season of a Hall of Fame caliber Major League Baseball player. The statistics were complete and comprehensive and allowed for many useful table comparisons between categorical variables. The results of the analysis dug fairly deep into the intricacies and mechanics of baseball and illustrated trends that may not have been evident to the common fan.

Barry Bonds tended to perform the best during the first half of games and with 0 or 1 out as opposed to there being 2 outs in an inning already. This would generally be expected for most players, but what separated Bonds was the number of times he was walked when he came to bat as a result of pitchers not wanting to pitch to him because of how accomplished a hitter he was.

The statistics showed a tendency for teams to walk him at the beginning and end of games for strategic reasons and since he is less of a threat when he doesn't have a bat in his hands.

7.2. Example 2-Linear Models

An Analysis of How Age and Weight Affect Blood Fat

Abstract

The purpose of this paper was to analyze a dataset that contained sample ages, weights and blood fat levels for 25 random patients to see if the blood fat level could be fitted to a linear model using age and weight as the input variables. The resulting model would be of the form:

$$y = a_0 x_0 + a_1 x_1 + a_2 x_2$$

where

a_0 = intercept,

a_1 = the weight of the patient,

a_2 = the age of the patient.

The model is created by using Linear Regression Analysis within the SAS Enterprise Guide.

Results showed that a statistically significant model could be created which proved that a linear relationship did in fact exist. However, one of the input variables was far less important than the other, which will be illustrated in analysis and graphs further along.

Introduction

Health professionals generally use two different measurements when it comes to assessing cardiovascular health – blood pressure and blood fat. Simple blood tests can be performed to determine the level of blood fat in a patient. High levels of blood fat can lead to heart issues and may impair the body's ability to carry cholesterol into and out of tissue. As such, it is important to examine how a patient's age and weight may affect their blood fat level to get a better understanding of the relationship if one exists.

This paper will attempt to attach a concrete linear model to the relationship in hopes of discovering whether or not age and weight are significant causes of high blood fat. If so, details on how they individually affect blood fat will be explored as well. This will be accomplished through the use of linear regression.

A statistically significant model was discovered in which the variability of blood fat levels was related to the variability of age and weight. As will be seen in later analyses, age played a much more important role in predicting blood fat levels than weight did.

Method

The dataset used in this analysis was obtained from a Linear Regression Datasets website (http://people.scs.fsu.edu/~burkardt/datasets/regression/regression.html) after being retrieved from D.G. Kleinbaum and L.L. Kupper's *Applied Regression Analysis and Other Multivariable Methods*. The data are complete and consist of information from 25 people ranging in ages from 20 to 60.

The data were saved to a text file and downloaded and converted into a SAS dataset for analysis. The variables were all numeric and continuous in nature, which leads to a fairly simple linear analysis. The results included scatter plots of Weight vs. Blood Fat and Age vs. Blood Fat as well as the output tables from linear regression.

Data Analysis

Since there were only 25 observations in this dataset, they are shown in their entirety in Table 1 below:

Table 1. Blood Fat Dataset

PatientNo	Weight	Age	BloodFat
1	84	46	354
2	73	20	190
3	65	52	405
4	70	30	263
5	76	57	451
6	69	25	302
7	63	28	288
8	72	36	385
9	79	57	402
10	75	44	365
11	27	24	209
12	89	31	290
13	65	52	346
14	57	23	254
15	59	60	395
16	69	48	434
17	60	34	220
18	79	51	374
19	75	50	308
20	82	34	220
21	59	46	311
22	67	23	181
23	85	37	274
24	55	40	303
25	63	30	244

The two input variables that exist in this dataset are Weight and Age. As such, they were used as the quantitative variables in the linear regression analysis. The goal was to see how they affected the dependent variable Blood Fat and whether or not a linear model could be used to represent the relationship. Figure 1 shows the resulting output.

Figure 1. Results of Linear Regression Analysis

Dependent Variable: BloodFat BloodFat

Source	DF	Sum of Squares	Mean Square	F Value	Pr > F
Model	2	102570.8147	51285.4073	26.36	<.0001
Error	22	42806.2253	1945.7375		
Corrected Total	24	145377.0400			

R-Square	Coeff Var	Root MSE	BloodFat Mean
0.705550	14.19623	44.11051	310.7200

Source	DF	Type I SS	Mean Square	F Value	Pr > F
Weight	1	10231.72620	10231.72620	5.26	0.0318
Age	1	92339.08845	92339.08845	47.46	<.0001

Source	DF	Type III SS	Mean Square	F Value	Pr > F
Weight	1	638.14894	638.14894	0.33	0.5727
Age	1	92339.08845	92339.08845	47.46	<.0001

Parameter	Estimate	Standard Error	t Value	Pr > |t|
Intercept	77.98253861	52.42963881	1.49	0.1511
Weight	0.41736210	0.72877608	0.57	0.5727
Age	5.21659081	0.75724446	6.89	<.0001

The topmost box gives the overall results of the model. Since the *p*-value was less than 0.0001, it is considered to be statistically significant meaning that the two input variables (Age and Weight) successfully explain the differences in the output variable (Blood Fat). The fairly high R-Square value of 0.705550 listed in the second box indicates that nearly 71% of the variability in Blood Fat is attributed to the variability in Age and Weight. This model does not describe the exact relationship between the input variables Age and Weight, and for the purposes of this analysis it will be assumed that the relationship is insignificant.

The third and fourth boxes show the Type I SS and Type III SS, respectively. Type III SS will be used since the result is invariant of the order of input variables. Interestingly enough, the *p*-values in this table show that Age is a statistically significant predictor but Weight is not, meaning that Age had enough of an impact on the Blood Fat level to dominate whatever relationship Blood Fat had with Weight since the overall model remained significant.

Figure 2 below provides a scatter plot of these two individual relationships. It should be clear that Age has a strong linear relationship with Blood Fat while Weight has nothing of the sort as previously discussed.

Figure 2. Graphs of Age and Weight vs. Blood Fat

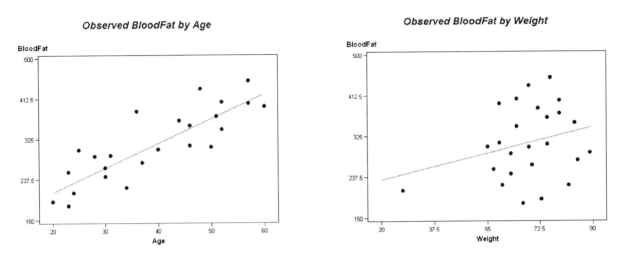

On the left graph, as the Age increased from 20 to 60, the Blood Fat level increased fairly steadily from ~200 to ~400. The points on the right graph are globular in nature and lack much of a linear fit.

The final box in Figure 1 presents the estimates for coefficients of the linear model along with error values.

Discussion

The dataset used in this analysis did not have a large number of observations, but the ones that were given were successfully fit to a statistically significant model in which the Blood Fat levels of patients were linearly related to their Age and Weight. The model was not very comprehensive however, since there was not a statistically significant relationship between Weight and Blood Fat by themselves.

In the medical field, this means that age is a much more reliable predictor of blood fat levels in patients than weight, at least for the group of patients used in this analysis. If a much larger number of observations were used, the previous statement could be made with more confidence and details on how exactly weight affects blood fat could be explored.

Printed in the United States
208817BV00001B/30/A